Seven Ideas That Shook the Universe
Supplementary Notes

Second Edition

Thomas E. Emmons
Kent State University

KENDALL/HUNT PUBLISHING COMPANY
4050 Westmark Drive   Dubuque, Iowa 52002

Copyright © 1990, 1999 by Thomas E. Emmons

ISBN 0-7872-6418-0

Library of Congress Catalog Card Number: 99-64973

Kendall/Hunt Publishing Company has the exclusive rights to reproduce this work,
to prepare derivative works from this work, to publicly distribute this work,
to publicly perform this work and to publicly display this work.

All rights reserved. No part of this publication may be reproduced,
stored in a retrieval system, or transmitted, in any form or by any
means, electronic, mechanical, photocopying, recording, or otherwise,
without the prior written permission of Kendall/Hunt Publishing Company.

Printed in the United States of America
10 9 8 7 6 5 4 3 2 1

TABLE OF CONTENTS

Chapter 1	Introduction
Chapter 2	Scientific Method and Space
Chapter 3	Time and Matter
Chapter 4	The Earth
Chapter 5	The Celestial Sphere
Chapter 6	Heliocentric Theory
Chapter 7	Planetarium
Chapter 8	The Planets and Kepler's Laws
Chapter 9	Motion
Chapter 10	Acceleration
Chapter 11	Velocity versus Time Graph
Chapter 12	Isotropic Property of Space and Forces
Chapter 13	Newton's Laws
Chapter 14	Math Examples of Newton's Laws
Chapter 15	Circular Motion
Chapter 16	Work and Energy
Chapter 17	Kinetic and Potential Energy
Chapter 18	Conservation of Energy
Chapter 19	Heat and Temperature
Chapter 20	Absolute Zero and Units of Heat
Chapter 21	Transfer of Heat
Chapter 22	Phases of Matter
Chapter 23	Pressure
Chapter 24	Nature of Liquids and Gases
Chapter 25	Introduction to Waves - Part I
Chapter 26	Introduction to Waves - Part II
Chapter 27	Electromagnetic Waves
Chapter 28	The Electromagnetic Spectrum
Chapter 29	Coherent Light and Laser
Chapter 30	Reflection, Transmission, and Absorption
Chapter 31	Phosphorescence, Fluorescence, and Doppler Effect
Chapter 32	Big Bang Theory and Refraction
Chapter 33	Diffraction, Polarization, and Interference
Chapter 34	Photoelectric Effect and Bohr Theory
Chapter 35	The Ultraviolet Catastrophe and Quantum Mechanics
Chapter 36	Matter Waves, The Heisenberg Uncertainty Principle
Chapter 37	Michelson Morley Experiment and Relativity
Chapter 38	Nuclear Physics and Conservation Laws

INTRODUCTION

THE SEVEN IDEAS

1. <u>We are not at the center of the universe.</u> (Copernican Astronomy) Today it is common knowledge that our Earth is not at the center of the universe, but is merely a planet revolving around a typical star we call the sun. In the days of Columbus, however, many people believed that the Earth was flat and celestial objects traveled around us.

We now also understand that our sun is not unique, but merely one of countless stars in the physical universe. It only appears so much larger and brighter because it is much nearer to us. While the naked eye could count about 2500 stars in our sky on a very clear evening, telescopes reveal billions. These stars are found in clusters or galaxies, ours being the Milky Way, a group of 100 - 200 billion stars. There are many billions (perhaps as many as 50 billion) of these galaxies, each containing billions of suns, many perhaps with planets. These discoveries were a great letdown to our ego as we realized that we are not uniquely located at the center of the universe.

2. <u>We can predict the future.</u> (Newtonian Mechanics) Applying mathematics to the laws of nature, it is possible to predict events. For example, if the mass of a car is known as well as the force exerted by the engine and all external forces such as friction, we can predict how the car will move. This means we can tell where the car will be in the future. Some have tried to carry this principle to extreme by suggesting the movement of all things, down to single atoms, can be predicted. If this were true, then every event in the physical universe could be predicted in advance. This would leave no room for chance events and everything would be determined.

3. <u>Energy</u> The word is often spoken, but very few people really understand its meaning. In the simplest terms, energy is what makes the universe go. Without energy, there would be no motion. The total amount of energy in the universe is a constant -- it cannot be changed. Whenever we use energy, we are withdrawing it from one source and applying it to another. The final form of energy, though, is in the form of oscillating motion of the atoms -- what we call heat.

4. <u>Heat death of the universe</u> As all forms of energy in the universe continue to be converted into heat, the atoms of the universe increase their vibrational motion. Every day that goes by, there is more heat. Eventually, unless there is some other catastrophic event, the universe will end up with nothing more than a conglomeration of small particles moving in chaotic patterns. This is sometimes referred to as the heat death or entropy death of the universe.

5. <u>There is an ultimate speed limit in the universe.</u> (Theory of special relativity) At this time, the fastest speed anyone has traveled relative to the Earth is about 25,000 mph, the speed obtained by

the astronauts on their return trip from the moon. Eventually we may go faster, but not faster than 186,000 miles per second, the speed of light. As we approach it, unusual things happen. First, size decreases while mass increases. If the speed of light were reached, the object would disappear and its mass would reach infinity. Also, as strange as it may sound, as the speed of light is approached, time slows down.

6. <u>Everything has a dual nature.</u> (Quantum Mechanics) We must look at everything from two points of view -- the wave aspect and the particle aspect. An important fact about waves is that they can pass through one another. Particles, on the other hand, act like solid objects and bounce off one another. Everything in nature seems to have both aspects. Objects usually thought of as solids must also be thought of as waves and vice versa.

Because the location of an object cannot be pinpointed within the size of the wave that goes along with its dual nature, we cannot predict the future of very small things such as atoms, we can only give probabilities. This discovery of the dual nature, or quantum mechanics, has been received with great enthusiasm by those worried about determinism -- the idea that all future events could be predicted. Quantum Mechanics has brought back the idea of chance or "free will" to the universe.

7. <u>Conservation Laws</u> There exist certain rules of nature which we cannot change called conservation laws. For instance, energy cannot be created or destroyed. This is known as the conservation of energy. Another example would be the conservation of charge. A positive or negative charge cannot be created or destroyed, but they can neutralize each other. Momentum, the mathematical product of an object's mass times its velocity, is another quantity that is conserved. We shall see the conservation laws serve as a foundation on which many theories of the physical universe are based.

The above seven ideas are not necessarily the only ones that should be considered earth shaking. Each scientist has his or her own opinions as to what is most important. If we add one more, number eight would be the discovery of the expanding universe -- the so-called Big Bang Theory. It appears that around 16 billion years ago all stars, planets, moons, etc. were compressed into one dimensionless point. Then a giant explosion occurred, throwing everything outward, including the Milky Way galaxy. Perhaps gravity will stop this outward rush and the physical universe will collapse only to repeat itself.

SCIENTIFIC METHOD AND SPACE

Physics is the study of the laws by which the physical universe operates. These laws are unique in that they were not made by man, nor can they be violated by man. No matter how rich or famous you are, you cannot escape the laws of nature. We have simply discovered them, mainly by scientists who use a five-step plan called the <u>scientific method</u>. This procedure works for all types of problems, both scientific and personal. The five steps are as follows:

1. Define the problem -- know your objective

2. Do research -- see if the problem has been solved before

3. Make hypotheses or guesses at possible solutions

4. Experiment -- test each hypothesis. For a personal problem, this can be a mental experiment based upon thinking through your ideas

5. State solution -- this is the hypothesis that has been proved and can now be stated as fact. Also give the percent of certainty, meaning how sure you are that your solution is correct.

The physical universe has three basic components. They are space, time, and matter. (The word mass can be interchanged with the word matter.)

SPACE

Space is that which separates objects. If there were no space, all objects would be lumped together in one very small spot. Our universe consists of three spatial dimensions; however, we can think how it would be in a world with a different number of dimensions. For instance, living in a one-dimensional world would be like being a point on a string with movement confined along the string. We can think of a two dimensional world as a flat surface. Trying to describe a four dimensional world is very difficult for us; however, it <u>can</u> be described mathematically.

[If the amount of space in a universe can be measured it is finite, but when space goes on forever, it is said to be infinite. A universe that has boundaries or limits is a closed universe, while one with no ending is an open universe. We do not know into which category our universe falls.] Imp't

UNITS OF SPACE

We use two systems when measuring space: the English and the metric. The English units (inch, foot, yard, mile) are based on human anatomy. For instance, a yard is the distance from the

2-1

nose to the outstretched fingertips. The metric system is based on our Earth. Scientists measured the Earth from the North Pole to the equator and divided this distance into 10 million segments, each being a <u>meter</u>. A meter is just a little longer than 39 inches.

As methods of measurements became more exact, the meter was redefined. Electricity was passed through the atoms of krypton gas, giving off an orange light. This light is emitted in evenly spaced ripples. The distance from one ripple to the next is the wavelength. The meter is now defined as 1,650,763.7321 wavelengths of orange krypton light.

Another easier way to define the meter is based on the principle that all light, whether it is from krypton or just from a standard light bulb, travels at the same speed. By turning a bulb on for an instant and allowing the light to travel for a certain length of time, the light will travel one meter. The length of time is 1/299,792,458 of a second.

Although the meter is the basic unit in the metric system, the meter can be subdivided into many other units. One hundredth of a meter is a centimeter, one thousandth of a meter is a millimeter. Meters can also be combined to make larger units. One thousand meters end to end is a kilometer or about 3/5 of a mile.

All of the above units are used when measuring space in one dimension such as the length of a room. When space is measured in two dimensions, such as the area of a floor, square units should be used -- square meters, square inches, square feet etc. Cubic units such as cubic meters, cubic inches, cubic feet, etc. are used when using all three dimensions to find volume.

TIME AND MATTER

TIME

Time is the second of the three entities that make up our universe. Time is defined as that which keeps events separate. It distinguishes the past from the present and the present from the future. The units of time are based on how long it takes a repeating physical event to occur.

One story tells that Galileo sat in church watching a swaying chandelier. Using his own pulse to keep track of time, he noted that regardless whether the chandelier went back and forth in a small arc or large arc, the length of time to oscillate was the same. He discovered the principle of the pendulum clock. What does affect it's time to swing is the <u>length</u> of the pendulum -- the shorter it is, the faster it will swing.

Another early method of measuring time was based on a spring coiling and uncoiling. Our first timepieces used this principle, while modern watches use a vibrating quartz crystal.

Unlike units of space, units of time are universal. Units under the English system are the same as those under the metric system. The day is based on the length of time it takes the Earth to rotate or spin on its axis which can be measured by timing how long it takes a star to appear to go once around the Earth. All celestial objects, including the sun, seem to move in a daily east to west direction as the Earth rotates <u>west to east.</u> When the sun sets in the west, it is actually a result of the western horizon coming up and blocking the sun.

The real length of time it takes the Earth to rotate is 23 hours, 56 minutes, and 4 seconds. This length of time is the <u>sidereal day</u>. We, however, live by the 24-hour span known as the <u>solar day</u>. This is because it takes the sun about four minutes longer to appear to go around the Earth -- a result of the movement of the Earth in its revolution about the sun.

The week is a unit of time not based on a physical event, but rather on superstition. The Babylonians noticed how a few points of light in the sky moved differently from the other stars. They saw seven that could move this way -- the sun, moon, and the five planets visible with the naked eye. They felt these seven must be powerful gods and therefore they worshiped each for a day. This gave us our 7-day week. Our week day names are based on this. Sunday is named for the sun, Monday for the moon, Mars (Tiu) for Tuesday, Mercury (Woden) for Wednesday, Jupiter (Thor) for Thursday, Venus (Freya) for Friday and Saturday received its name from the planet Saturn.

Our month was determined by the length of time it takes the moon to complete a cycle of phases. The actual time from one full moon to the next full moon is 29 days, 12 hours, 44 minutes, and 2.9 seconds.

One revolution of the Earth around the sun takes about 365 1/4 days (365 days 6 hours 9 minutes and 9.8 seconds). This is what determines the true year. Leap year comes every four years as the 1/4 days accumulate to make one full day.

As the Earth spins each day it wobbles very slowly; in fact, it takes 25,800 years to complete one such wobble. This is defined as a Platonic year. One effect this has on our Earth is that our north star changes.

The Babylonians, thinking that the whole number of months in a year was a lucky number, divided the daytime and nighttime portions of a solar day each into 12 segments resulting in a 24-hour day. Then they divided the hour into sixty minutes and the minute into sixty seconds. To the Babylonians, sixty was a nice round number just as one hundred is a nice round number to us. A new legally recognized unit of time is the "jiffy" which is 1/60 of a second.

MATTER

Matter, like space, has different units under the metric and English systems. A <u>kilogram</u>, a prime unit in the metric system, weighs 2.2 pounds on Earth. We must say "on Earth" because mass and weight are <u>not</u> the same. <u>Weight</u> depends on how strong gravity pulls on an object, while mass depends on the atoms that make up the object. A kilogram on Earth weighs 2.2 pounds, on the moon less than 1 pound, and much more on Jupiter. This demonstrates that mass is constant, but weight varies as a function of gravity. A much smaller unit of mass is the gram which is 1/1000 of a kilogram. It takes 453.6 grams to weigh 1 pound on Earth.

The English system has only one fundamental unit of mass, the <u>slug</u> that weighs 32.2 pounds on Earth. A person who weighs 161 pounds on Earth has a mass of 5 slugs.

There are two methods used to measure matter. One involves weighing the object using the familiar scales. For example, 2.2 pounds of weight on Earth corresponds to a mass of one kilogram; 4.4 pounds would be 2 kilograms; 6.6 pounds would be 3 kilograms etc. Likewise, a weight on Earth of 32.2 pounds would correspond to 1 slug; 64.4 pounds would be 2 slugs etc.

Although weighing an object is a common method to determine mass, it is useless in the microscopic world where one deals with things the size of atoms. It is also useless in the macroscopic world when one deals with the mass of stars and planets. To determine the mass of very large or small objects, we use a law of nature that states that the more mass in an object, the harder it is to start it moving and the harder it is to stop. Just as an empty grocery cart is easier to maneuver than one filled with 20 gallons of milk, an atom that does not contain much mass is easier to move than one that does. A force is exerted on an atom and its path is observed. If the atom gains speed rapidly, it is an atom with relatively little mass. The same principle applies to measuring the mass of very large objects.

THE EARTH

If seen from space our Earth appears as a round sphere even though it does bulge slightly at the equator and is somewhat flattened at the poles. The two points we refer to as the north and south poles are the ends of the axis around which our Earth rotates once each sidereal day.

The Earth's equator, about 25,000 miles around, divides the Earth into the northern and southern hemispheres. This circle around the Earth is a great circle because it divides the planet into two equal halves. There are other small circles north and south of the equator which would not cut the Earth into equal portions. These east-west small circles are lines of latitude. The equator, as a great circle, serves as the origin for latitude -- it is zero degrees. We measure from the equator to the north pole in positive numbers with the north pole being +90 degrees. The southern hemisphere is measured in negative numbers with -90 degrees representing the south pole.

To give the Earth's surface a coordinate grid, there are also north-south lines of reference. These lines, called lines of longitude, are all great circles; therefore, we must arbitrarily choose a starting point for the origin. This chosen place, a line that runs through Greenwich, England, is the prime meridian and is assigned a longitude of zero degrees. The International Date Line is close to 180 degrees east or west of the prime meridian.

The diameter of our planet is about 8,000 miles, making it 4,000 miles from the surface to the center of the Earth. Gravity tries to pull all objects to the Earth's center, explaining why people don't fall off the other side.

On the inside, the Earth consists of three layers. Outermost is the crust, a 20-mile thick layer of light rocks. Next is the mantle which is 1800 miles thick and consists of heavy rocks. The innermost layer, or core, is made mostly of the heavy metals iron and nickel. This layer is 2100 miles thick and exists in two states. At the very center of the Earth the materials are solid; however, around this solid is molten nickel and iron. This was determined by carefully timing vibrations from earthquakes as they travel from one side of the Earth to the other.

Just as the inside of the Earth is divided into three layers, the atmosphere also has three sections. Closest to our planet is the troposphere, a layer extending upward for 7 miles. This layer contains most of our air and is where most weather occurs. The stratosphere extends from 7 to 45 miles, is very cold, and has air too thin to support life. The final layer, the ionosphere extends from 45 to 200 miles. This is a gas that has lost electrons and is therefore positively charged. This acts as a mirror from which to reflect radio waves.

THE CELESTIAL SPHERE

Just as we use the lines of latitude and longitude to find an exact point on the Earth's surface, we need a coordinate system in the sky to locate celestial objects. There are two ways of doing this.

The easier method is based on the observer's sky and involves drawing a picture of the sky overhead. (See Diagram 5A) The circle at the bottom where the sky seems to touch the ground is the <u>horizon</u>. The observer is assumed to be at the center of this circle. The circle is divided into four sections, each being 90 degrees apart on the circle and representing the four directions or cardinal points. The point directly overhead is the <u>zenith</u>. If a line is drawn from the north, through the zenith, to the south, it would cut the sky into an eastern and western half. This line is the <u>meridian</u>. North on the horizon is 0 degrees with East being 90 degrees, South 180 degrees, and West 270 degrees. These degrees are used by planes and ships for navigation.

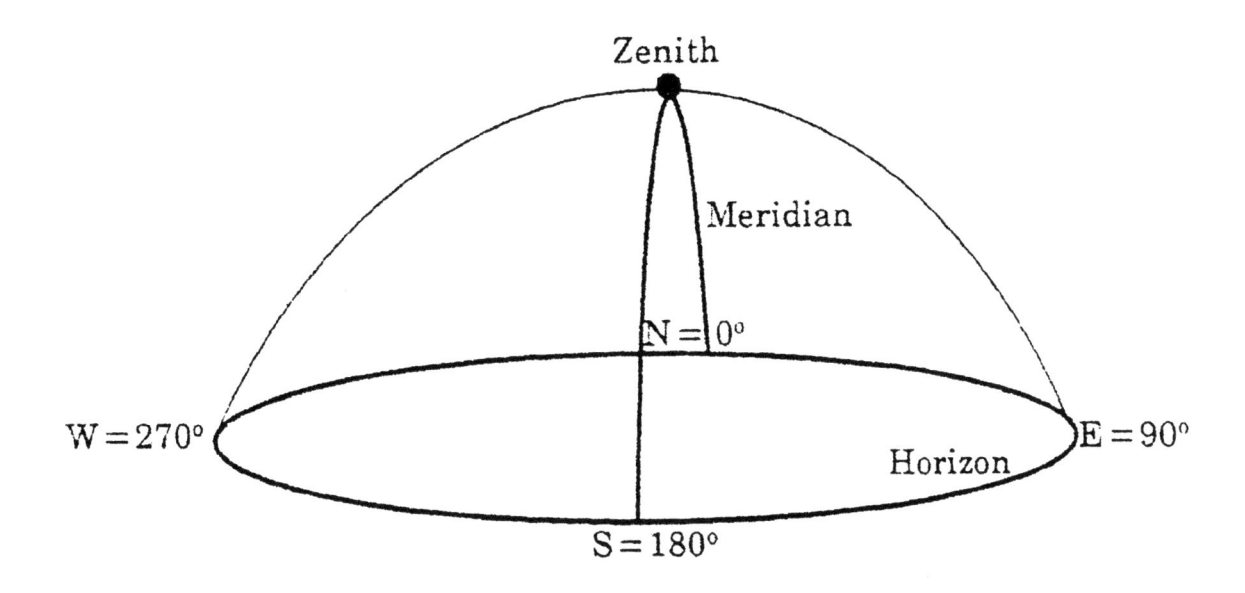

DIAGRAM 5-A

Using this method to find an object in the sky, one would need to know in what direction to look and how high above the horizon. The direction, referred to as the <u>azimuth</u>, is found using the 0 - 360 degrees around the horizon. For example, one would look in the east for an object with an azimuth of 90 degrees. <u>Altitude</u> tells how high the object will be in the sky. This ranges from 0 degrees on the horizon to 90 degrees at the zenith. For instance, if Jupiter has an azimuth of 135 degrees and an altitude of 30 degrees, it would be found 1/3 of the way above the horizon in the southeast. While easy to use, this method has one serious drawback -- it is a local system.

These coordinates are not the same for an object when viewed from different places on Earth.

To solve this problem, we need a universal system. Here we may make use of the <u>geocentric theory</u> which assumes the Earth is located at the center of an imaginary transparent sphere called the <u>celestial sphere</u>. The imaginary points directly above the north and south poles are the north celestial pole (NCP) and the south celestial pole (SCP). The circle around the sky over the equator is termed the <u>celestial equator</u>. Polaris, commonly called the north star, is very close to the north celestial pole and is unique because of its location. The only place on Earth where Polaris is seen directly overhead is at the north pole.

It is important to know how to locate Polaris. First, we find the familiar Big Dipper in the sky. Using the two outside stars in the bowl of the Big Dipper as pointer stars, we can locate Polaris which is the last star in the handle of the Little Dipper. (See Diagram 5B)

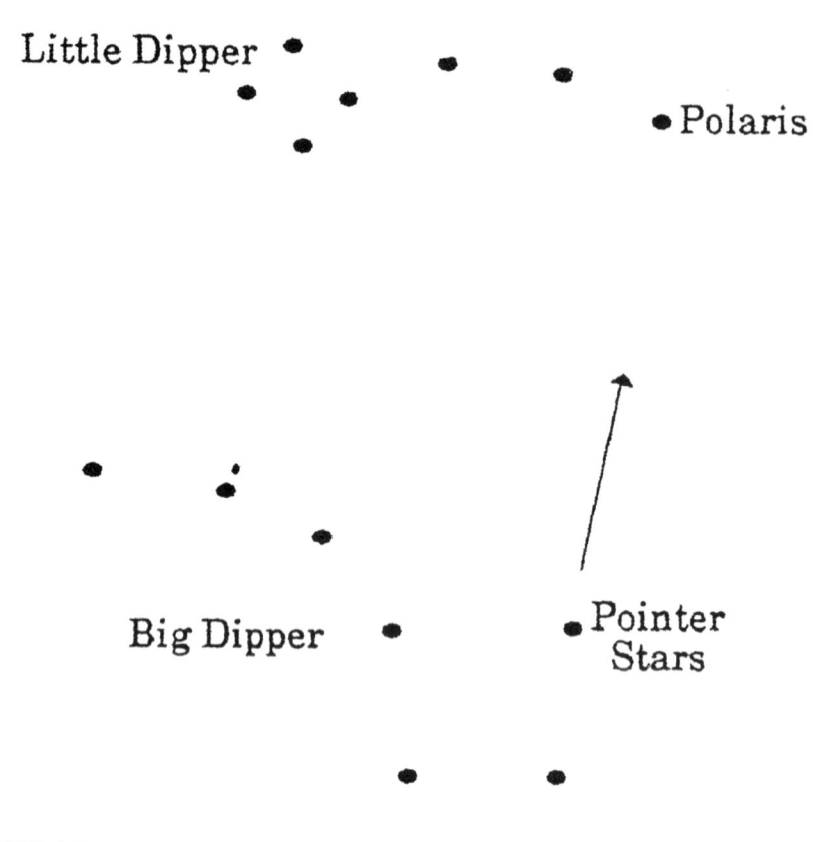

DIAGRAM 5-B

The north pole of the Earth is at a latitude of 90 degrees. When Polaris is viewed from the north pole, it is seen at an altitude of 90 degrees. The equator has a latitude of 0 degrees, and when Polaris is viewed from the equator, it has an altitude of 0 degrees. This is no coincidence, but one of two rules pertaining to the north star Polaris.

Rule 1 states that the altitude of Polaris in the sky is equal to the observer's latitude. Rule 2 states that as the Earth rotates, Polaris does not appear to move, but all other celestial objects appear to travel in circles around Polaris once each day.

Viewed from the north pole, Polaris would be over your head. Once each sidereal day stars close to Polaris would appear to trace small circles around it, while those farther away would make larger circles. Stars at the north pole do not rise or set and are known as <u>circumpolar stars</u>. At the equator all stars rise and set. In the middle latitudes, such as here in Kent, some rise and set while others are circumpolar.

Also projected onto the celestial sphere are the lines of latitude and longitude. (See Diagram 5C) Latitude in the sky is called <u>declination</u> and is measured from 0 degrees at the celestial equator to plus or minus 90 degrees at the poles. North-south lines when projected onto the celestial sphere are called <u>right ascension lines</u>. On Earth, 0 longitude is at Greenwich, England; however, the line above Greenwich cannot be used as 0 right ascension because this line is constantly changing as the Earth rotates. Instead, the north-south line in the sky that runs through the apparent position of the sun on the first day of spring (vernal equinox) was chosen as the starting point for right ascension.

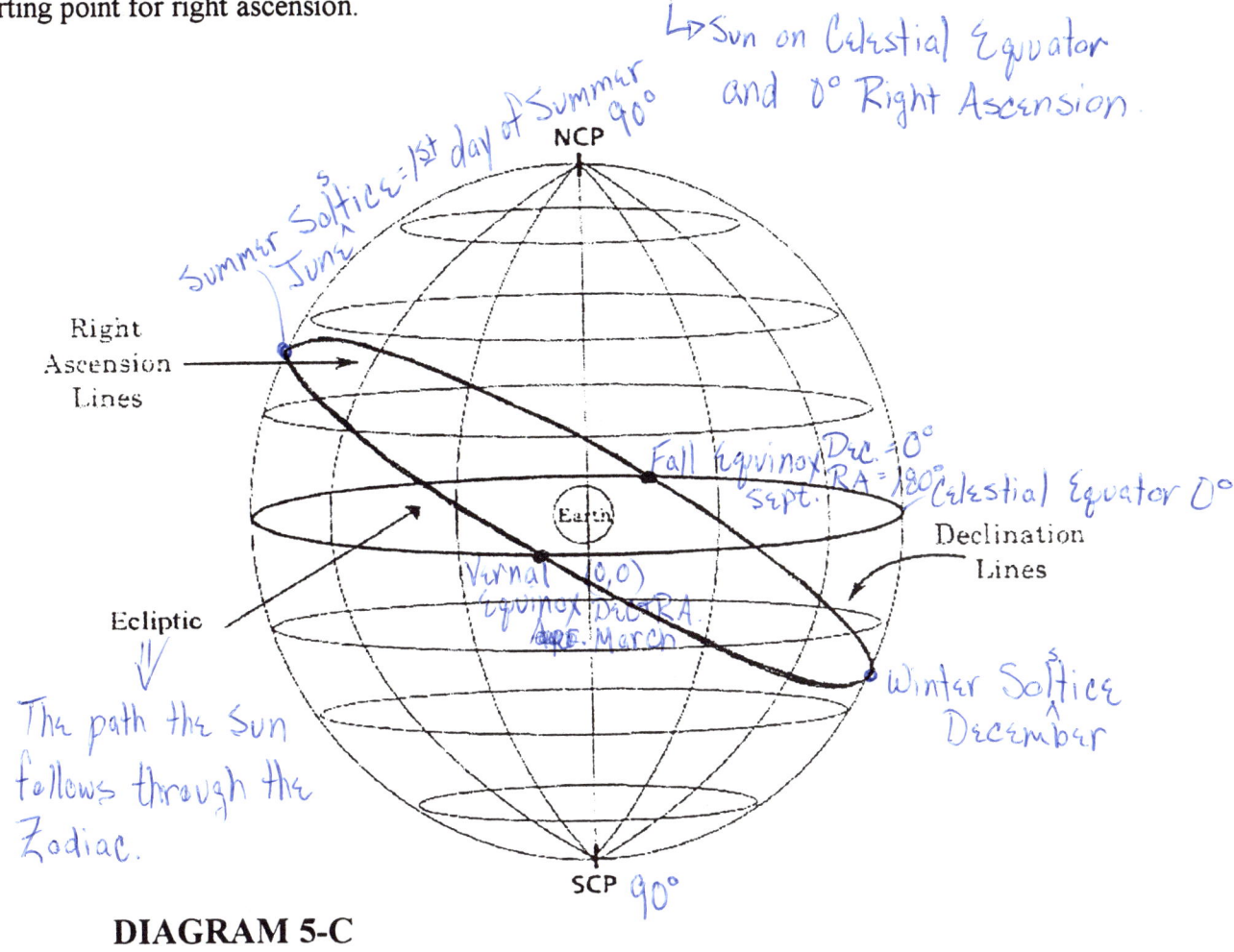

Sun on Celestial Equator and 0° Right Ascension

Summer Solstice = 1st day of Summer — June — 90°

Fall equinox Dec = 0° RA = 180° Sept.

Vernal (0,0) equinox Dec = 0° RA = 0° March

Winter Solstice December

Ecliptic — The path the Sun follows through the Zodiac.

NCP 90°
SCP 90°

DIAGRAM 5-C

Zodiac = 12 constellations the Sun moves through.

5-3

HELIOCENTRIC THEORY → Sun Centered

Geocentric Theory - Aristotle = Earth Center Theory

In the 15th century most people still believed in the geocentric or Earth-centered theory. Those who had contradicted it had been persecuted. Nicholas Copernicus (1473-1543) however persisted in his belief that the sun, and <u>not</u> the Earth, was at the center of the solar system. This concept is the called the <u>heliocentric theory</u>. We still do not know the exact center of our universe, but we do know that in our own solar system, the planets revolve around the sun. People became convinced of this heliocentric theory after the following seven arguments were presented.

1. <u>Retrograde motion of the planets</u> = *Planets are to crisscross in the sky*

For centuries man had noted that the planets would cross the sky, stop, and reverse their direction. Copernicus explained that these planets only <u>seem</u> to move in retrograde or backward motion because the Earth can pass a planet as it journeys around the sun. This is not unlike a car appearing to be traveling backwards as you pass it on an expressway.

2. <u>Nova of 1572</u> = *New*

At this time a nova, or exploding star, appeared. It was bright enough to be seen in the daytime. This in itself was not a proof of the heliocentric theory, but it did get people thinking about the sky and also questioning their preconceived idea that the stars on the celestial sphere would never change.

3. **Moons of Jupiter** — *Proved Objects Orbit other Objects. Planet means Wanderer. (1564-1642)*

While observing Jupiter through his newly invented telescope, Galileo noticed that the planet's moons did not remain stationary. He saw that nature likes to put a large mass in the middle with smaller objects revolving around it. Jupiter and its moons represented a mini solar system to Galileo.

4. **Phases of Venus**

Planets, emitting no light of their own, shine only by reflected sunlight; therefore, only the side of the planet toward the sun is light. When Galileo watched Venus, he observed it going through phases. (See Diagram 6A)

6-1

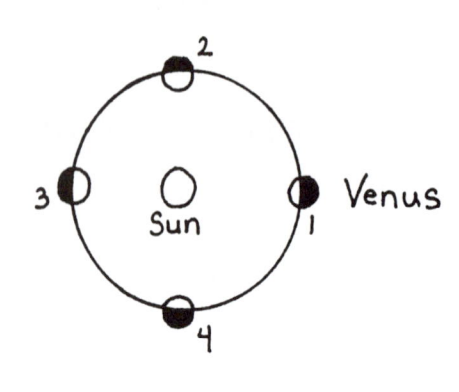

DIAGRAM 6-A

When Venus is in Position 1, Galileo would see just half of a circle; in Position 2, a whole circle; in Position 3, a half circle again; and in Position 4, he would not be able to see the planet at all because the dark side was facing him. This proved that the sun was in the center of at least the orbit of Venus.

5. <u>Parallax of stars</u>

The farther an object is away from an observer, the longer it takes to pass him or her. If the heliocentric theory is true, stars close to the Earth should go by more quickly than more distant ones. Astronomers observed this effect by noting close stars would shift their position against the more distant stars as the Earth moved from one side of its orbit to the other, thus proving the Earth goes around the sun. (See diagram 6B)

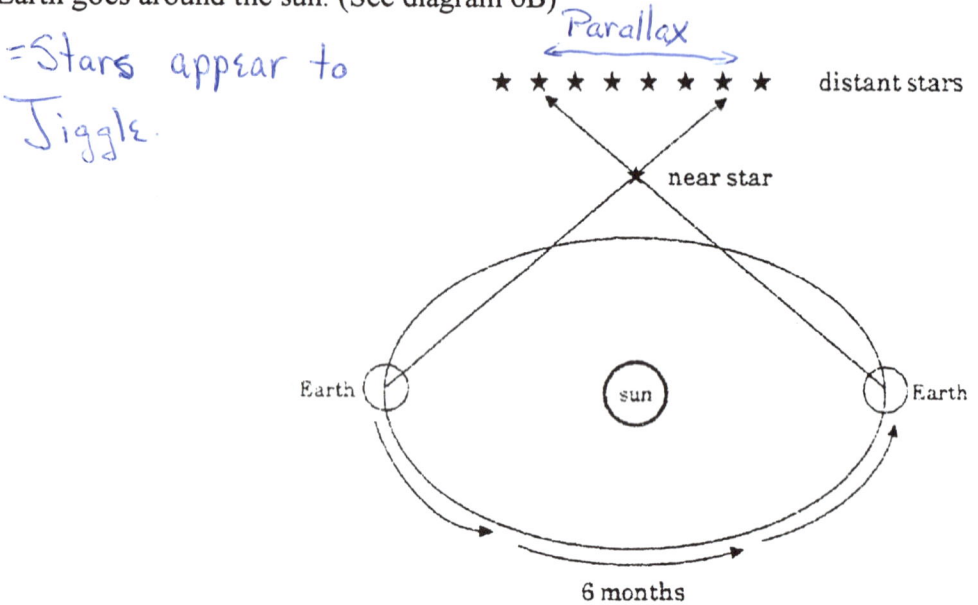

= Stars appear to Jiggle.

DIAGRAM 6-B

6. <u>Seasons</u>

The seasons we experience on Earth are <u>not</u> determined by our distance from the sun. In fact, we are closest to the sun in January and farthest from it in July. We have seasons because the Earth's axis is tilted 23 1/2 degrees away from the perpendicular to the plane of its orbit and because <u>the Earth goes around the sun.</u> (See Diagram 6C)

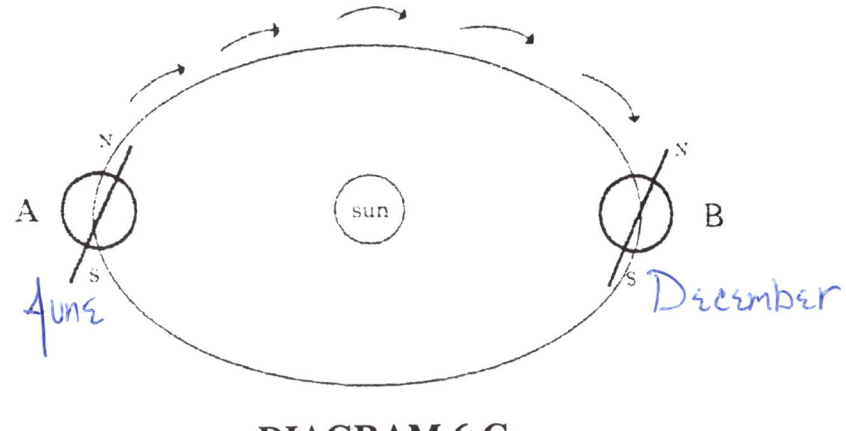

DIAGRAM 6-C

June (A) ... *December* (B)

When the Earth is in Position A, the northern end of the Earth is pointed toward the sun; thus, it is receiving more sunlight and experiencing warmer weather. Because the Earth travels around the sun once each year, in six months the Earth will move from Position A to Position B. Now the northern portion of the Earth is pointed away from the sun and receiving less sunlight; therefore, it is having the colder weather of winter. Seasons are reversed below the equator.

7. <u>Aberration of starlight</u> — *Most Convincing*

This is probably the most convincing argument to substantiate the heliocentric theory. This is similar to the optical illusion of snowflake paths seeming to bend into the front windshield of a moving car. In the same way, light rays radiating from a star seem to bend into the front of the moving Earth. When the Earth is on the other side of its orbit, the light <u>seems</u> to bend in the opposite direction. This proves the Earth must be moving. (See Diagram 6D)

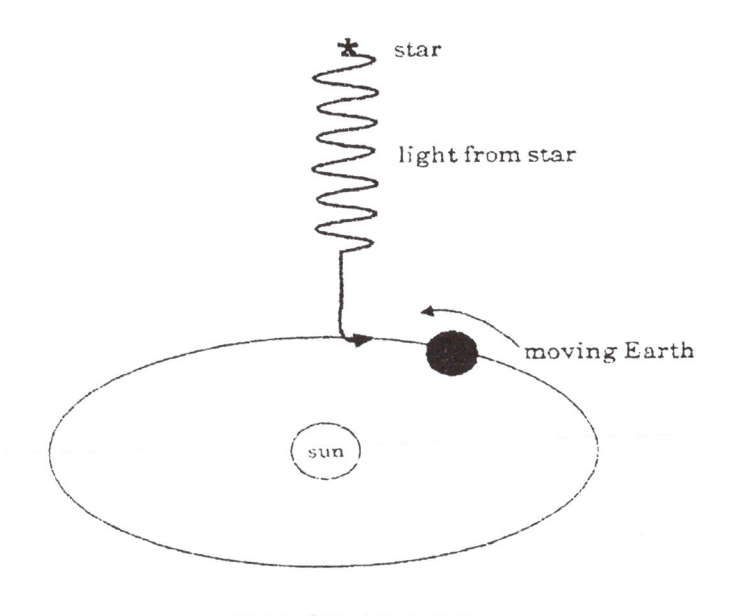

DIAGRAM 6-D

PLANETARIUM

The planetarium is a place where one can see a simulation of the nighttime sky. An instrument in the center of the room projects stars, planets, the sun, and moon onto a large dome. A large ball on top of the planetarium projector contains approximately 1500 accurately positioned holes. A bright xenon light inside shines through these holes producing points of light on the dome corresponding to the stars in nature's sky.

Below the star ball, seven analog computers calculate and position the seven celestial objects that can move against the background stars. These are the sun, moon, Mercury, Venus, Mars, Jupiter, and Saturn.

At the bottom of the projector and around the base of the pedestal, there are several auxiliary projectors that can show lines in the sky to represent right ascension lines and declination lines, the ecliptic, the meridian, and many others.

In addition to the obvious advantage of being able to see the stars at any time in the planetarium, there are several other benefits. By using the planetarium instrument, the effect of the Earth's rotation can be shown speeded hundreds of times. We can watch the movement of all the celestial objects around the Earth in just a few minutes. This would take a day in nature's sky. We can also observe the effect of the Earth's revolution around the sun speeded thousands of times. What would take one year to accomplish in the real sky can be seen in only a few minutes in the planetarium.

Still another advantage of the planetarium is the ability to show the sky as it appears from different points on the Earth. In a matter of a few seconds, the planetarium projector can be shifted to show the sky as seen from the north pole, the equator, or any other location on Earth.

THE PLANETS & KEPLER'S LAWS

(Kepler (1571-1630)
Brahe (1546-1601)

According to current theory, our solar system was formed from a huge cloud of gas about 4 1/2 billion years ago. All the planets revolve around the sun in about the same plane except Pluto, suggesting that it may have been a later addition to our solar system in the form of a giant captured rock. Our sun with its planets is only one of the 100 billion stars in the Milky Way Galaxy. The nearest galaxy of any consequence, the only one visible with the naked eye in the Northern Hemisphere, is the Andromeda Galaxy.

Nearest the sun is the planet Mercury that is similar to our moon in appearance. Mercury has no atmosphere with surface temperatures that can reach over 800 degrees Fahrenheit.

The thick atmosphere surrounding Venus acts as an insulator to keep the surface temperature as high as 900 degrees Fahrenheit. Due to this extreme heat, its surface under the clouds looks like dry, cracked mud.

The next planet from the sun, of course, is the Earth and fourth is Mars. The red appearance of Mars is due to red surface dust. White polar caps composed of frozen carbon dioxide and water are visible in the north and south. Mars has several very large craters, probably from volcanoes, and a large crack in the surface that would dwarf the Grand Canyon.

Next comes the largest planet Jupiter which spins very rapidly causing its equator to bulge. The famous red spot is a whirlpool on its surface composed of methane and ammonia gas. A moon of Jupiter called Io is the only place we know of where there are active volcanoes other than on Earth.

Beyond Jupiter is Saturn distinguished by the beautiful system of rings surrounding it. Composed of particles covered by ice, there are actually many rings which were possibly formed from a moon that fragmented.

More has been discovered about Uranus since Voyager flew by it in January of 1986. One interesting thing about Uranus is that its axis is tipped nearly 90 degrees compared with the Earth's 23-degree tilt. There are few surface features on this planet. It appears as a greenish-blue ball.

In August of 1989 the Voyager spacecraft passed Neptune and sent back pictures showing a blue planet with whispers of clouds composed of droplets made of liquid methane. Pluto, discovered in 1930, is our farthest known planet. It might have been a former moon of Neptune that started revolving on its own around the sun or a very distant rock traveling through space captured by the sun's gravity.

Moons revolve around some of the planets. Mercury and Venus have none. Jupiter and Saturn have the most. Our own moon is one of the seven largest in the solar system. It is less dense than Earth. Its composition is similar to the outer layer of the Earth, but has an ash gray surface of basalt. It is covered with craters resulting from meteoroids striking its surface. Some current theories suggest that the moon spun off from the Earth in the early days of our solar system.

Much information about the solar system has been obtained from telescopes such as the one at Mt. Palomar; however, the best pictures have come from spacecraft such as Voyager.

An early astronomer, Tycho Brahe (1546-1601), observed the planets, collecting a great deal of information about their movement in the sky. Later Johannes Kepler (1571- 1630) studied this data carefully, arriving at three laws governing how the planets travel around the sun.

1. The first law states that planets travel around the sun in ellipses with the sun at one of the foci, not at the center; the other focus is empty. A planet, therefore, is not always the same distance from the sun. The Earth, for example, is not always 93 million miles from the sun. It is closest to the sun around January 4th and farthest away around July 4th.

2. While studying Tycho Brahe's observations, Kepler noted that when the Earth is close to the sun (Diagram 8A, Position A) it covers a greater distance in its orbit in a designated time than it does in the same length of time when far from the sun (Diagram 8A, Position B). He found that a line joining a planet and the sun (radius vector) sweeps out equal areas in equal times. In the diagram the shaded area near Position A will be equal to the shaded area near Position B. This means the Earth had to move faster when in Position A so that its short radius vector could sweep out the same area as the long radius vector at Position B. In other words, Kepler's second law requires that a planet travel faster when close to the sun and slower when far from the sun.

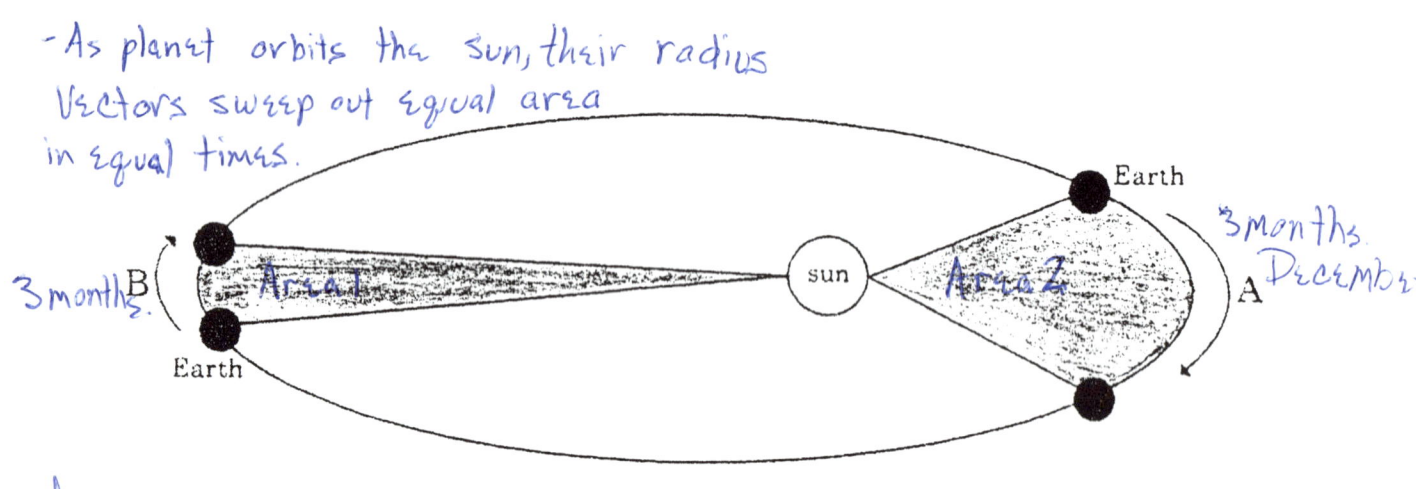

- As planet orbits the sun, their radius vectors sweep out equal area in equal times.

3 months B ... *Area 1* ... sun ... *Area 2* ... *3 months December A*

DIAGRAM 8-A

Area 1 = Area 2
Earth moves 21 mi/sec in Jan. + 19 mi/sec in July.

3. Kepler's third law is a mathematical equation which can be used to determine the distance between a planet and the sun. By applying his equation, we see that the larger the planet's orbit, the longer it takes to go around the sun. Mercury which has the smallest orbit takes only three months to travel around the sun, while Pluto having the largest orbit, takes about 250 years.

Don't Need to know Properties of Planets, but need to know the order.

MOTION

The second of the seven ideas deals with our ability to predict the future under certain circumstances. If we know where an object is and what forces are applied to it, we can predict where it will be after a given interval or increment of time. This requires the study of various types of motion.

The first type of motion, uniform motion, deals with the movement of an object at a constant speed in a constant direction. To calculate uniform velocity, the change in an object's distance is divided by the change in time. Using the Greek letter delta (Δ) to mean change, the above can be written in abbreviated form as the following equation:

$$V = \frac{\Delta d}{\Delta t}$$

V represents the uniform velocity

Δd represents the distance moved

Δt represents the increment of time

EXAMPLE

A straight road between two cities is 140 miles long. A car which is 20 miles down the road from the first city drives to the second city between 1 P.M. and 4 P.M. Assuming the car moved at a constant speed, what is its uniform velocity?

$$\Delta d = 140 \text{ miles} - 20 \text{ miles} = 120 \text{ miles}$$
$$\Delta t = 4 \text{ hours} - 1 \text{ hour} = 3 \text{ hours}$$

$$V = \frac{\Delta d}{\Delta t}$$

$$V = \frac{120 \text{ miles}}{3 \text{ hrs}}$$

$$V = 40 \frac{miles}{hr}$$

EXAMPLE

Sound travels at the approximate speed of 1100 feet per second. How long would it take thunder to travel 5500 feet which is just a little more than one mile?

NOTE: The way the equation was originally written will not work for this problem since we already know the velocity. In this case, Δt is the unknown. Using the tools of algebra, the original equation can be rewritten to solve for Δt.

Starting with $V = \dfrac{\Delta d}{\Delta t}$ **and then solving for Δt gives:**

$$\Delta t = \frac{\Delta d}{V}$$

$$\Delta t = \frac{5500 \; feet}{1100 \; \dfrac{feet}{sec}}$$

$$\Delta t = 5 \; sec$$

NOTE: The unit we were left with (seconds) is appropriate because we were solving for time. This is not just a coincidence, but a basic and extremely useful rule in physics. It can be used as a check for correct problem solving. If the units do not make sense, an error has been made!

Another type of motion, <u>relative velocity</u>, occurs when a person (the observer) watches an object move inside a moving box (frame). This would be similar to an observer on the curb watching a passenger walk in the aisle of a moving bus. The passenger would be the object and the bus would be the frame.

To determine the speed at which an object moves by the observer outside the frame, the velocity of the object is added to the velocity of the frame. Written in abbreviated form, the equation is:

$$V_R = V_O + V_F$$

V_R represents the velocity relative to outside observer

V_O represents the velocity of object within the frame

V_F represents the velocity of frame as seen by observer

EXAMPLE

A person sees a bus moving 20 feet per second. A passenger in the bus walks from the back of the bus to the front with a speed of 5 feet per second. How fast does the observer see the passenger move by him?

$$V_R = V_O + V_F$$
$$V_R = 5 \text{ feet/second} + 20 \text{ feet/second}$$
$$V_R = 25 \text{ feet/second}$$

EXAMPLE

In this same bus, a passenger seated in the front throws a ball toward the back of the bus with a speed of 20 feet per second. Relative to the observer outside the bus, how fast does the ball move?
NOTE: By convention, objects moving to the right or upward are considered to have a positive velocity, while objects moving to the left or downward have a negative velocity. If the bus is moving to the right, its velocity would be positive 20 feet per second. Since the ball is thrown backward (to the left), its velocity would be negative 20 feet per second.

$$V_R = V_O + V_F$$
$$V_R = -20 \text{ feet/second} + 20 \text{ feet/second}$$
$$V_R = 0 \text{ feet/second}$$

In the above example, relative to the observer, the ball does not move right or left; however, to people inside the bus (frame), the ball appears to be moving to the rear of the bus at 20 feet per second.

The effect of relative velocity works fine with every day speeds; however, when the speed of light is approached, the formula fails. A bullet fired with a speed of 110,000 miles per second from the front of a spaceship that is moving 100,000 miles per second does not move past a stationary observer with the speed of 210,000 miles per second as the formula would suggest.

Although it seems to defy common sense, the observer would see the bullet moving at a speed of less than 186,282 miles per second. This is a result of a law of nature that says that no matter what is done, nothing can move faster than 186,282 miles per second (the speed of light) relative to any observer.

ACCELERATION

Acceleration is defined as a change in an object's velocity divided by the change in time. This can be written as the following equation:

$$a = \frac{\Delta V}{\Delta t}$$

a represents acceleration

ΔV represents the increment of velocity

Δt represents the increment of time

EXAMPLE

A spacecraft is launched at 9:43:10. At 9:43:40 its speed is 120 feet per second. What is its acceleration?

$$\Delta V = 120 \text{ feet/sec} - 0 \text{ feet/sec} = 120 \text{ feet/sec}$$
$$\Delta t = 9:43:40 - 9:43:10 = 30 \text{ sec}$$

$$a = \frac{\Delta V}{\Delta t}$$

$$a = \frac{\frac{120 \text{ feet}}{\text{sec}}}{30 \text{ sec}}$$

$$a = \frac{4 \text{ feet}}{\text{sec}^2}$$

Notice the units in which acceleration is expressed. This means that for every second that goes by, the spacecraft will increase its velocity by 4 feet per second. The following chart shows the craft's speed at one-second intervals:

Δt	VELOCITY
0 sec	0 feet/sec
1 sec	4 feet/sec
2 sec	8 feet/sec
3 sec	12 feet/sec
4 sec	16 feet/sec
5 sec	20 feet/sec

The next chart shows the velocity for an object which is moving 30 feet per second to begin with and then accelerates at -6 feet per second, per second (or -6 ft/second squared).

Δt	VELOCITY
0 sec	30 feet/sec
1 sec	24 feet/sec
2 sec	18 feet/sec
3 sec	12 feet/sec
4 sec	6 feet/sec
5 sec	0 feet/sec

The minus sign tells us that the object is slowing down or decelerating. For every second that goes by, the object reduces its speed by six feet per second.

An important rate of acceleration is 32 feet per second squared. This is the acceleration of a falling object near the Earth's surface due to gravity. This acceleration is true for all objects, regardless of shape, weight, or size *if* we disregard the effect of the air. The chart below indicates what happens to the speed of a falling object here on Earth if air resistance is neglected. Every second that goes by, the object gains 32 feet per second in its speed.

Δt	VELOCITY
0 sec	0 feet/sec
1 sec	32 feet/sec
2 sec	64 feet/sec
3 sec	96 feet/sec
4 sec	128 feet/sec
5 sec	160 feet/sec
6 sec	192 feet/sec

The <u>terminal velocity</u> is the maximum speed an object can reach before air resistance prevents it from continuing to accelerate. Feathers, leaves, pieces of paper reach a terminal velocity very quickly. For a person, the terminal velocity is much higher. It is approximately 140 miles per hour, but does vary with the position of the body. Extending the arms and legs would slow one down, while folding into a ball would increase the terminal velocity.

There is an old saying that if you drop a penny from the top of the Empire State Building and hit someone with it, that you could kill him. This would be true if there were no atmosphere. Since the air does slow down the fall of the penny, you can drop the coin over the side of the building and not worry about being a murderer! In fact, if we had no atmosphere, even raindrops or snowflakes would come down to Earth at an amazing speed. We would have to dodge them during a storm!

It is important to remember that <u>without the effect of the air, all objects accelerate through the entire fall at 32 feet per second squared.</u> This means that all objects dropped from the same height would reach the ground at the same time.

VELOCITY ≈ TIME GRAPH

The equation $V = \Delta d / \Delta t$ (equation for uniform velocity) could be solved algebraically for Δd, i.e. $\Delta d = V \Delta t$. However, when using this equation to solve for distance, there must be <u>no</u> acceleration. The object must not be gaining or losing velocity! A new equation is required when seeking the distance an object moves when it is accelerating. This new equation is:

$$\Delta d = 1/2\ a\ \Delta t^2$$

Δd represents the distance moved

a represents the acceleration

Δt represents the interval of time

It is important to remember when using this formula that it is assumed that the object is starting from rest; that is, it has no beginning or initial velocity.

EXAMPLE

A car starts out at a green light with an acceleration of 1.5 meters per second squared. If the car continues to accelerate, how far will it move in 7 seconds?

$$\Delta d = 1/2\ a\ \Delta t^2$$

$$\Delta d = 1/2\ \frac{1.5\ meters}{sec^2}\ 49\ sec^2$$

Note: 7 sec X 7 sec = 49 sec^2

$$\Delta d = 36.75\ meters$$

Dropped objects accelerate at 32 feet per second squared here on Earth, neglecting air resistance. Because a dropped object starts out at rest and accelerates, the equation $\Delta d = 1/2 \, a \, \Delta t^2$ can be used to determine how far an object falls.

EXAMPLE

A diver walks off a diving board which he thinks is 75 feet high. Using a water-proof stop watch, he notes that it takes three seconds before he hits the water. Is the board really 75 feet high?

$$\Delta d = 1/2 \, a \, \Delta t^2$$

$$\Delta d = 1/2 \, \frac{32 \; feet}{sec^2} \, 9 \; sec^2$$

Note: 3 sec X 3 sec = 9 sec^2

$$\Delta d = 144 \; feet$$

No, the board is not 75 feet high. It is 144 feet high.

In real life, motion usually involves a variety of accelerations. When traveling across town, one goes through many variations in speed. To find the distance traveled when going through complex motions, a graph becomes very useful. The horizontal scale depicts the passing of time, while the vertical scale shows the velocity of an object; therefore, the graph is called the <u>velocity versus time graph.</u> If an object is accelerating, the plotted graph will slant upwards; if an object is decelerating, the graph will slant downwards, and if the object is moving with uniform motion, the plot is a straight horizontal line.

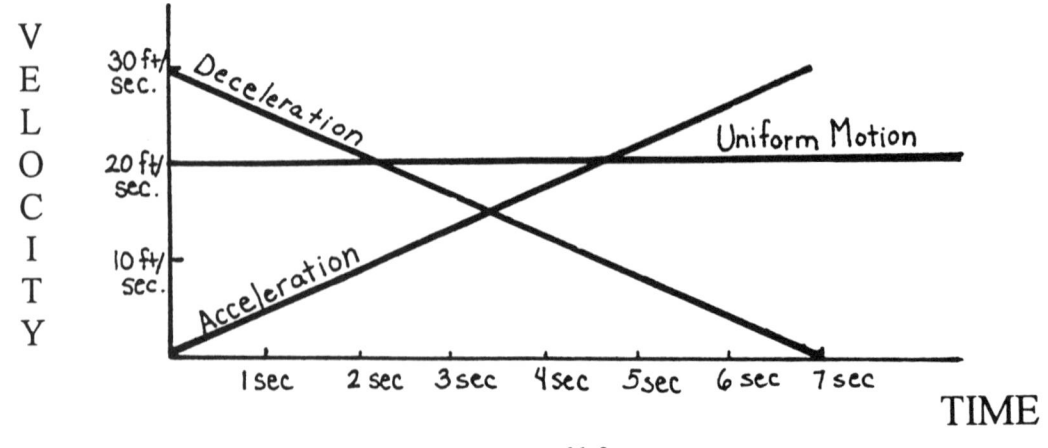

For complex motion, the graph might look as follows:

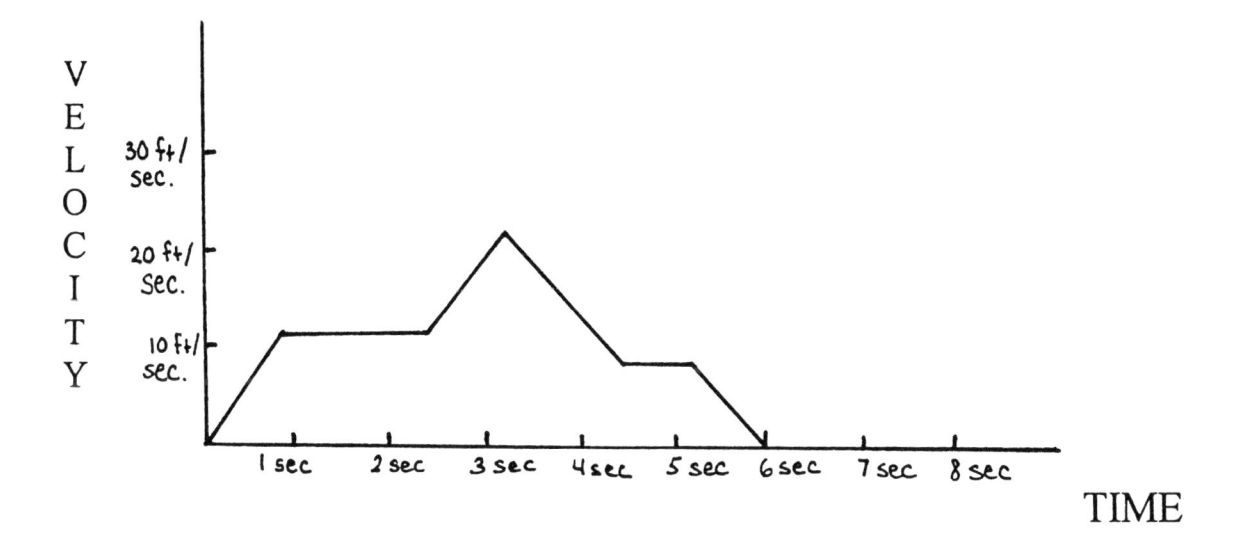

Just by glancing at the graph, acceleration at any time can be determined by looking at the slope of the line. At the time of one second, the object is accelerating; at two seconds, it is moving with uniform motion; at three seconds, it is accelerating; at four seconds, it is decelerating; and at five seconds, it is moving with uniform speed, etc.

If the slope of the line cannot be determined by looking at the graph, there is a branch of mathematics called <u>differential calculus</u> that will give the slope mathematically.

There is another benefit to be obtained from the velocity versus time graph in addition to finding an object's acceleration. The area under the curve represents the distance traveled. One way to approximate the area is to divide it into many rectangles as shown below.

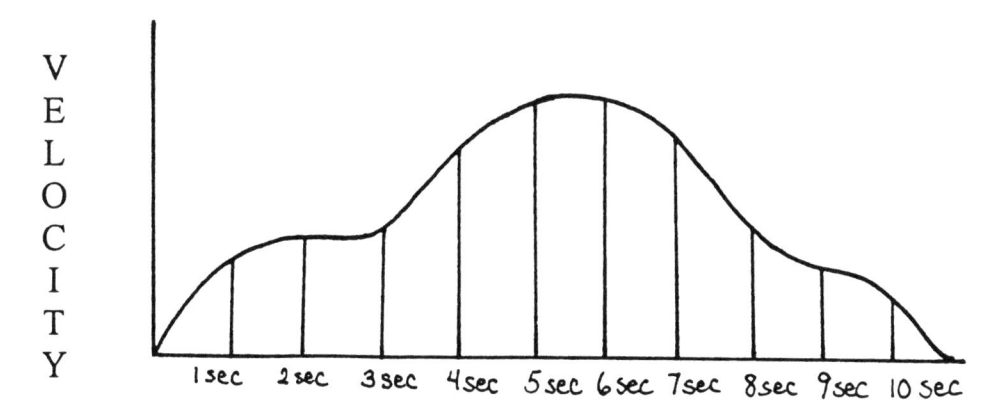

The area of any one rectangle would be the product of its width and height. Adding all the rectangles together would give the total area. This would represent the distance moved by the object. If more rectangles are used, the area can be found more precisely, but this requires a great deal of tedious work.

Fortunately, there is a branch of mathematics that will divide the area into an infinite number of rectangles, and add the areas together to give the total area. This is <u>integral calculus</u>.

Not all areas under the curve require calculus; instead, the total area can be found by first using simple formulas to find the area of small rectangles and triangles and then summing these areas. The area of a rectangle is found by multiplying its base (b) times its height (h). A triangle's area is found by multiplying 1/2 of its base (b) times its height (h).

EXAMPLE

A car starts out at a red light and accelerates to a speed of 30 feet per second in two seconds. The car then moves with the uniform speed of 30 feet per second for two seconds. The brakes are then applied and the car comes to a stop in one second. How far did the car travel in the total five seconds?

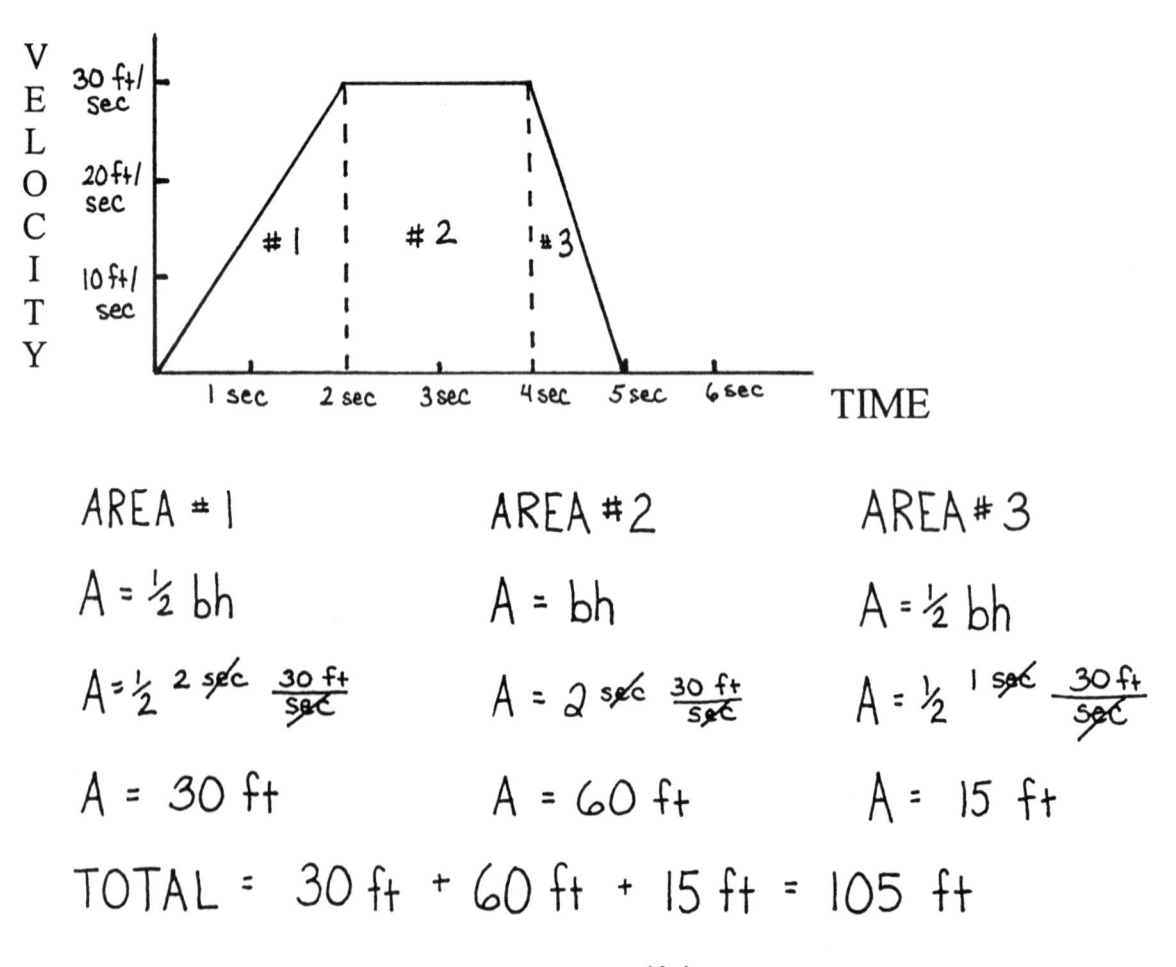

[Graph: Velocity (ft/sec) vs Time (sec). Trapezoidal shape divided into region #1 (triangle, 0–2 sec), region #2 (rectangle, 2–4 sec), region #3 (triangle, 4–5 sec). Peak velocity 30 ft/sec.]

AREA #1

$A = \frac{1}{2} bh$

$A = \frac{1}{2} \cdot 2 \, sec \cdot \frac{30 \, ft}{sec}$

$A = 30 \, ft$

AREA #2

$A = bh$

$A = 2 \, sec \cdot \frac{30 \, ft}{sec}$

$A = 60 \, ft$

AREA #3

$A = \frac{1}{2} bh$

$A = \frac{1}{2} \cdot 1 \, sec \cdot \frac{30 \, ft}{sec}$

$A = 15 \, ft$

TOTAL = 30 ft + 60 ft + 15 ft = 105 ft

ISOTROPIC PROPERTY OF SPACE AND FORCES

An object can move in more than one dimension at the same time, such as moving along a curved path. However, when moving in two or more dimensions simultaneously, motion in one dimension does not affect motion in other dimensions. This is due to the isotropic property of space.

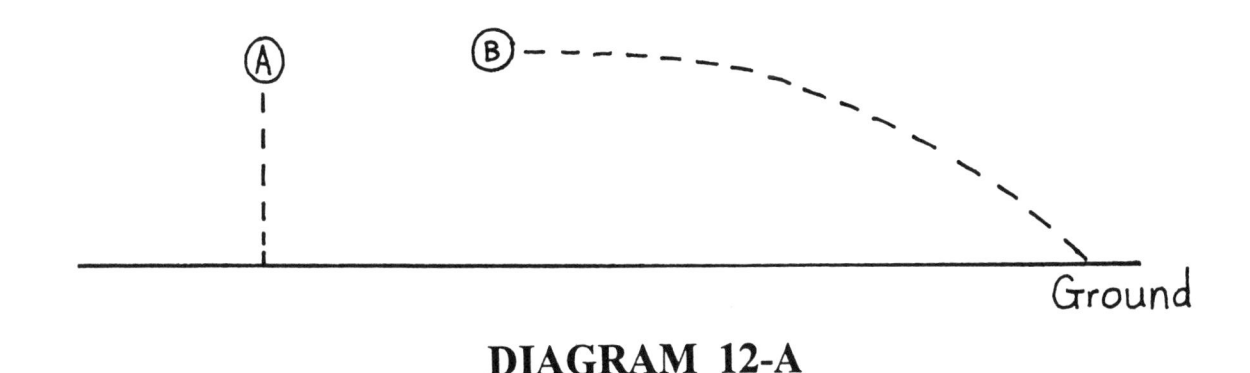

DIAGRAM 12-A

In diagram 12A two stones are released from identical heights at the same instant, one being dropped and the other thrown horizontally. Gravity will pull down on both with an acceleration of 32 feet per second squared. In other words, the fact that stone B is thrown sideways will in no way influence how gravity will pull it in the vertical direction. Both stones will strike the level ground at the same time!

Likewise, a bullet shot from a horizontal gun and another dropped from the same height will both hit the ground simultaneously. It makes no difference at what speed the bullet leaves the gun. If the effect of the air is neglected and we assume the gun is fired horizontally over level ground, the bullets will hit the ground at the same time.

In the previous chapters we have discussed several types of motion. The next logical topic to consider is what causes motion? In order to start motion, a force, defined simply as a push or a pull, is needed. All the forces of nature, and there are many of them, can be put into the following four classifications:

1. Gravity

This force is produced by all matter, regardless of its size; and yet, it is the weakest of the four forces. In every day life we don't notice the effect of gravity produced by an object unless the object is very large, such as a celestial object. Gravity is an attracting force, one that pulls things together.

2. Electromagnetic

This force is produced by particles with either a positive or negative charge. It follows the law "Like charges repel and unlike attract." Two positives or two negatives fly apart, while a negative and positive attract one another. What would be the electromagnetic force between a positive proton and a neutral neutron? Here the electromagnetic force would be zero because an electromagnetic force is produced only by charged particles, not neutral ones. In this case, only the force of gravity would be experienced.

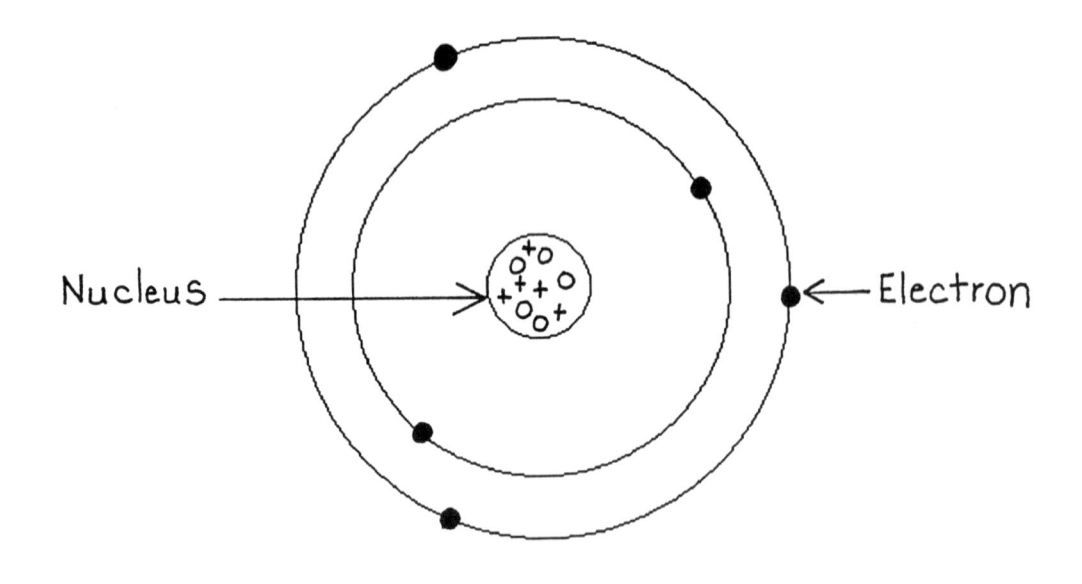

DIAGRAM 12-B

An atom, as shown in diagram 12B, is electrically neutral although it does have positive and negative charges. There are usually the same number of positive protons as negative electrons and their charges cancel. The electrons are found in outer shells. The center (nucleus) is composed of protons as well as neutral particles called neutrons. The number of protons in the nucleus determines the element. These protons, all being positively charged, repel one another and would not stay together in the nucleus without the next force.

3. Strong Nuclear Force or Strong Interacting Force

This force, nicknamed the nuclear force and the strongest of the four forces, holds the nucleus of an atom together. Particles called mesons act like a "glue" to hold the nucleus together.

4. Weak Nuclear Force or Weak Interacting Force

A neutron in the nucleus of an atom will live indefinitely if embedded there; however, it becomes unstable when taken out of the nucleus. In this situation, it will live for only about 19 minutes. Then it will break apart into a proton, electron, and neutrino. The fourth force, the weak interacting force, causes this break-up of a free neutron. A neutron alone causes great damage. This is the principle for the neutron bomb. Here great amounts of neutrons are released which harm living tissue in people or animals, but allow buildings etc. to remain untouched. After approximately 19 minutes, the neutrons break into the three harmless parts and the area would be safe to enter.

Many physicists are working on an idea called the GUT theory (grand unification theory) which is an attempt to unify all forces into just one. Partial success has already been achieved by combining the weak interacting force and the electromagnetic force. Modern nuclear research has found indications that the strong nuclear force may also be combined with the weak nuclear force and electromagnetic force. Combining gravity seems to be the present stumbling block to the grand unification theory, but it is thought that this hurdle will eventually be overcome.

NEWTON'S LAWS

After studying motion, Sir Isaac Newton (1642-1727) formulated some well-known laws pertaining to the relationship between force and motion. He first described what happens in nature when a force is <u>not</u> exerted on an object. This first law of Newton can be called the "No Force Law." The law states that an object which is not moving will remain stationary. It also states if the object is moving, it will remain moving with a constant speed (uniform motion) in a straight line. We have trouble visualizing this because most objects we see come to rest due to friction. If we could remove friction, the object would remain moving. Another way to say this is that an object wants to remain in the same state -- if it is stationary, it will remain stationary; if it is moving, it will remain moving.

Newton's second law describes what happens when a force <u>is</u> put on an object. This must be a special force called a <u>net</u> force, meaning the amount of force remaining after you subtract forces in the opposite direction. For example, if one person is pushing with a force of 30 pounds on an object while another person is pushing in the opposite direction with a force of 10 pounds, the <u>net</u> force would be 20 pounds. When you are sitting in a seat, gravity is pulling down with a force equal to your weight, but this is not a net force because the seat is pushing up canceling out your weight.

Newton's second law states that a net force will cause an object to accelerate. When a train pulls out of a station, the engine exerts a net force and it accelerates. As long as the engine continues to exert a net force, the train will continue to go faster and faster. As the train does go faster, the force of friction gets larger and larger. This extra friction pushes back on the train, diminishing the net force of the engine. Eventually the train will reach a speed where the force of friction will completely cancel the force of the engine, meaning that the net force is zero. At this point, even though the engine is working very hard and consuming great quantities of fuel, its only accomplishment is to counteract friction. Although the net force of the engine is zero, if it were turned off, the force of friction would become the net force and the train would decelerate to a stop. If the force of friction could be eliminated, we would only use fuel to accelerate the train to a required speed. We could then turn the engine off and coast forever.

Newton's second law has two corollaries. The first says that the greater the force on an object, the greater the acceleration. The second states that the greater the mass, the smaller the acceleration. These corollaries seem to agree with everyday experiences except in the case where we drop a light object and a heavy object from the same height at the same time. Applying the idea that the more mass an object has, the smaller the acceleration, we would expect the massive object to start out very slowly like a ten-ton truck at a stoplight. This would mean that the light object should hit the ground first; however, it can easily be demonstrated that they hit the ground at the same time.

The explanation for this apparent paradox is that <u>there is a greater pull of gravity on the more massive object</u>! Neglecting the effect of the air, gravity causes all objects to accelerate at the same rate, but this is accomplished by gravity pulling more in proportion to the object's mass.

Newton's second law can be put into mathematical form as follows:

$$a = F/m \quad \text{OR} \quad F = ma$$

a = acceleration

F = net force

m = mass

The three units by which force is measured are derived from Newton's second law. The three result as a product of a mass multiplied by an acceleration.

If a mass of one slug is made to accelerate one foot per second squared, then the net force on the object is defined as one pound.

If a mass of one kilogram is made to accelerate one meter per second squared, then the net force on the object is defined as one newton.

If a mass of one gram is made to accelerate one centimeter per second squared, then the net force on the object is defined as one dyne.

Newton's third law says that you cannot put a force on an object without it exerting a force back on you. You cannot lean on a wall without it pushing back on you. We see this exemplified when a person jumps out of a small rowboat. The person is pushed forward while the boat is pushed in an opposite direction. The forward force on the person is the action and the opposite force on the boat is the reaction. Consequently, Newton's third law could also be stated, for every force of action there is an opposite and equal force of reaction.

Newton later added a fourth law pertaining just to the force of gravity. He determined that there is an attracting force between any two masses which could be determined by multiplying the amount of mass in one object times the amount of mass in the other object and then dividing this product by the square of the distance between the centers of the two objects.

MATH EXAMPLES OF NEWTON'S SECOND LAW

[START]

The equation a = F/M, Newton's Second Law, is perhaps the most often used equation in physics and engineering. It can be used to predict the future. If the mass of an object is known and the value of the net force placed on an object is also known, we can calculate the object's acceleration. And once the acceleration is determined, the distance the object moves can be obtained by using $\Delta d = 1/2 \; a \; \Delta t^2$.

Example

In 7 seconds, how far will a toy wagon move on a frictionless level surface if a girl pushes on it with a force of 20 pounds? The mass of the wagon is .5 slug. (It would weigh 16 pounds on Earth.)

First Find Acceleration

$$a = \frac{F}{M}$$

$$a = \frac{20 \; slug \; \frac{feet}{sec^2}}{.5 \; slug}$$

NOTE: $1 \; pound = slug \; \frac{foot}{sec^2}$

$$a = 40 \; \frac{feet}{sec^2}$$

14-1

Next Compute Distance Moved

$$\Delta d = 1/2 \ a \ \Delta t^2$$

$$\Delta d = 1/2 \ \frac{40 \ feet}{sec^2} \ 49 \ sec^2$$

$$\Delta d = 980 \ feet$$

Example

Referring to the previous example, what would be the distance moved if the wagon were not on a frictionless surface, but rather it moved over a dirt road where friction pushed against the wagon with 6 pounds of force? Newton says it is the _net_ force that causes acceleration and the net force would now be the 20 pounds provided by the girl minus the 6 pounds of friction. Net force = 14 pounds.

To Find Acceleration

$$a = \frac{F}{M}$$

$$a = \frac{14 \ slug \ \frac{feet}{sec^2}}{.5 \ slug} \qquad NOTE: 1 \ pound = slug \ \frac{foot}{sec^2}$$

$$a = 28 \ \frac{feet}{sec^2}$$

To Find Distance Moved

$$\Delta d = 1/2 \ a \ \Delta t^2$$

$$\Delta d = 1/2 \ \frac{28 \ feet}{sec^2} \ 49 \ sec^2$$

$$\Delta d = 686 \ feet$$

Example

Besides predicting the future location of objects, Newton's second law can be used to quantitatively determine the mass of an object. If a net force of 30 newtons is applied to a stone and it accelerates 6 meters per second squared, what is the mass of the stone in kilograms (kg)?

$$M = \frac{F}{a}$$

$$M = \frac{30 \ \frac{kg \ meter}{sec^2}}{6 \ \frac{meter}{sec^2}} \qquad NOTE: \ 1 \ newton = kg \ \frac{meter}{sec^2}$$

$$M = 5 \ kg$$

CIRCULAR MOTION

If a car is seen to be increasing its speed, we say it is accelerating. If the car is decreasing its speed, it is decelerating. What about a car that maintains a constant speed, but changes direction? Even though it doesn't move faster or slower, it is still accelerating. The reason for this goes back to the isotropic property of space. The three dimensions of space are independent of one another.

When an object moves around a curve, it is in effect slowing down in one dimension while gaining speed in another dimension. Thus the object is undergoing two accelerations at once -- an increase in speed in one direction while decreasing speed in another.

Referring to diagram 15A, when the ball is at the top of the curve (position #1), it is moving horizontally toward the right. But when the ball is at the side of the curve (position #2), the ball is moving in a downward direction. It stopped moving sideways and started moving downward. It has undergone acceleration.

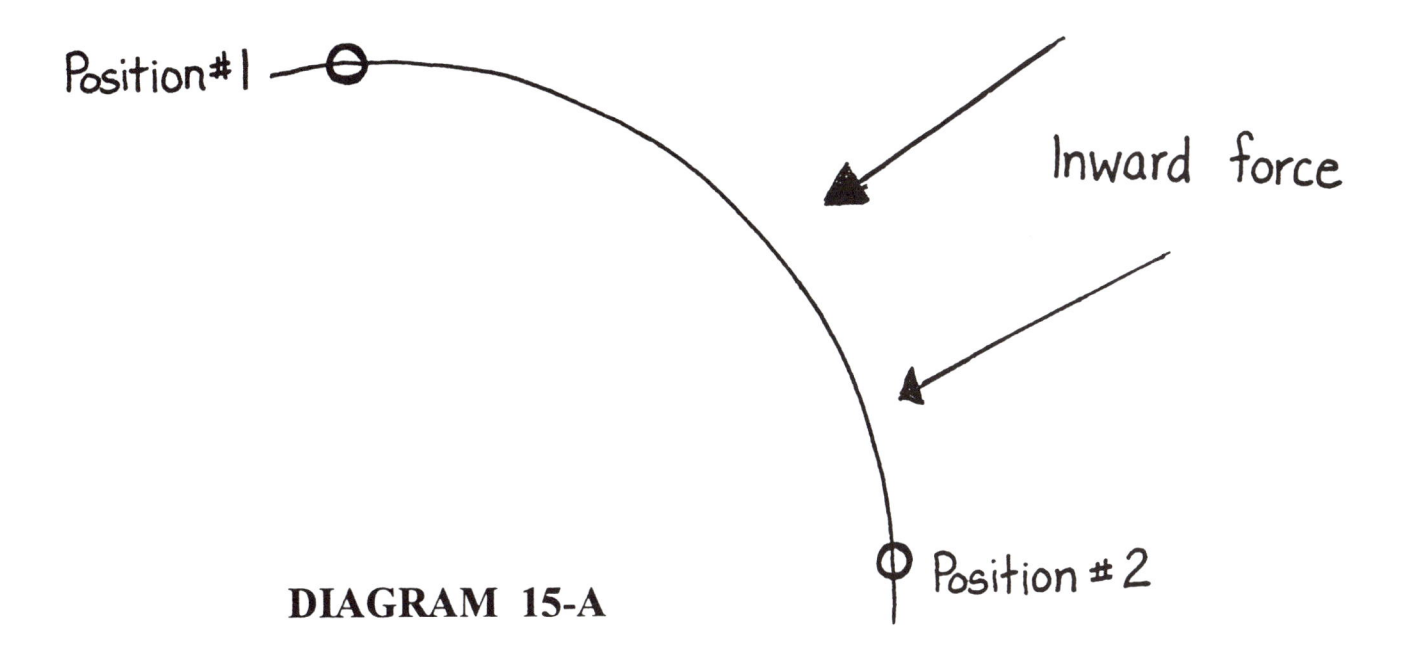

DIAGRAM 15-A

Newton says that in order for an object to accelerate, there must be a net force acting on it. To make an object move faster, the force pushes forward on the object. To make an object slow down, the force must push back on the object. To make an object accelerate by changing its direction, the force pushes sideways. It pushes the object toward the center of the arc around which it is moving. The arrows in diagram 15A show the direction of the force needed to make the ball round the curve. This inward force is centripetal force. It can come from the force of gravity, the electromagnetic force, the strong interacting force, or the weak interacting force.

Newton's third law tells us that for every force pushing in one direction, there will be an equal but opposite reacting force. The centripetal force pushing an object inward around a circle will generate a reacting force which is felt by the object. This outward force is centrifugal force that should really be called centrifugal reaction.

This outward centrifugal force is felt by people rounding a curve in an automobile. We are forced outward away from the center of the arc as friction between the front tires and the road pushes the car inward. You might think of centrifugal force as a force you feel when a car rounds a bend. Inside this car, you tend to keep going straight and therefore you push against the door while the car is turning.

On the scale of a single atom, we see the roll played between centripetal force and centrifugal force. The positive protons in the nucleus pull inward on the negative electrons (opposite charges attract). The electrons would fall into the nucleus if it were not for the fact that they are moving. The inward pull on the electrons is a centripetal force causing the electrons to move in an orbit, while the outward centrifugal reaction holds the electrons away from the nucleus.

On a large scale, the sun's gravity pulls inward on the Earth. We would fall into the sun if we were not moving in our orbit. The sun's gravity supplies the centripetal force and the Earth experiences centrifugal force holding us away from the sun.

When a satellite is sent into orbit around the Earth, it does not remain in orbit because it is away from the Earth's gravity. The Earth's gravity still pulls the satellite toward the center of the Earth, but because the satellite was launched so as to attain a speed of 18,000 miles per hour parallel to the surface of the Earth, gravity accelerates the satellite by changing its direction. This causes the satellite to experience centrifugal force and holds it above the ground.

Any object moving in an orbit has a quantity called angular momentum. Angular momentum will cause an object to move faster if the radius of the orbit is decreased. Likewise, if the radius is increased, the object will slow down. We see this exemplified when a spinning ice skater pulls his or her arms into the body, causing the skater to spin faster.

WORK AND ENERGY

Energy is defined as the ability to do work. But what is work? In ordinary terms, we usually think of work as something that makes us tired; however, the scientific definition of work does not agree with our everyday connotation. To the scientist, three requirements must be met in order to accomplish work.

1. A force must be applied to an object.

2. The object must move through a distance.

3. The direction of the force and the direction of displacement must be in the same dimension.

If a person just satisfies the first requirement, he or she could become very tired, and yet no work is done. Holding a 50 pound sack of flour over one's head for two hours would be very tiring, but no work is accomplished. For work to be done, not only does one have to apply a force to an object, but the object must be moved as well.

What about carrying a sack of flour across a room? Strange as it may sound this too does not involve work. This is because the force is upward and the distance moved is sideways. The force and the movement must be in the same dimension. (Note: Technically some work was done when the flour was initially pushed sideways to start it moving.) What about raising a sack of flour off the floor? Lifting a sack of flour does constitute real work because when lifting, a force is applied in the vertical direction and the flour is moved in the vertical direction.

Since energy is the ability to do work, it can now be defined as the ability to push on an object and move it in the same dimension as the force. Energy is like having a bank account, except instead of having money to spend for material goods, it is something to use to accomplish work.

The units by which work and energy are measured involve a unit of force and a unit of distance. If an object is pushed a distance of one foot with a force of one pound, then one foot pound of work is done. If an object is pushed with a force of one newton (about 1/5 pound) and moved a distance of one meter, one could say a newton meter of work is done. The newton meter has been condensed into the word joule. The last unit is the force of a dyne applied to an object that is moved one centimeter. This is the dyne centimeter that has been condensed into the word erg.

The word powerful is often confused with energetic; however, the terms have two different meanings. Power involves time. Power is the work done divided by the amount of time required to do the work. A mouse could do just as much work as a gorilla if the mouse were given enough time. They both could be described as energetic; but since the gorilla could do the work in a much shorter period of time, it is more powerful.

Units for power are derived from a unit of work divided by a unit of time. One foot pound of work done in one second is a foot pound/second. A joule of work done in one second is called a watt. 746 joules of work done in one second is a horsepower.

KINETIC AND POTENTIAL ENERGY

The concept of energy as a savings account is very useful; however, instead of purchasing items with our energy, we use it to do work. Just as there are different kinds of bank accounts, there are different kinds of energy.

Kinetic energy is the energy found in all moving objects and thus anything that is moving has the ability to do work. A fly buzzing around a room, a car moving down a street, a train rolling along a track, an airplane flying in the air, or a galaxy drifting through space all have kinetic energy.

To spend this energy and produce work, a collision is required. When a moving object makes impact with a second object, the second object is pushed some distance by the first one. Thus, work is done.

The amount of kinetic energy in a moving object is found by taking 1/2 of its mass and multiplying it by the velocity squared. Because the velocity is squared, a small increase in speed will result in a large increase in kinetic energy. For instance, if the velocity of an object is doubled, it has four times the kinetic energy (two squared is four). If the velocity of an object is tripled, the kinetic energy increases by a factor of nine (three squared is nine).

One might want to keep this in mind when deciding to travel at 60 miles per hour rather than 30 miles per hour. There will be four times as much damage and injury at 60 miles per hour should an accident happen.

Another type of energy is potential energy of which there are several types. One is due to position. In this type, the location of an object can give it the ability to do work. For example, a rock on the top of a cliff has the potential to do work for us. All we have to do is allow the rock to fall. When it gets to the bottom, it could push with a force on a wooden stake, driving it some distance into the ground.

Water on top of a dam would be another example of potential energy due to position. The water could flow over the dam, pushing on the blades of a turbine which could turn a generator to produce electricity.

A second type of potential energy exists in a stretched or compressed spring. When the coils of a spring are displaced from their normal positions, they will try to return to their original positions. An object attached to the end of the spring will experience a force, moving it some distance when the coils are released.

Charged particles can also do work and thus are a form of potential energy. Moving charged particles through a motor can cause the motor to exert a force and move an object some distance. Usually we use electrons as the charged particles because their mass is small and thus easy to move. Electricity is nothing more than a stream of moving electrons.

Chemical potential energy is the energy released during a chemical reaction. Gasoline, for example, contains potential energy. Gasoline is chemically combined with oxygen when it is burned in the cylinder of an automobile. This burning produces gas which pushes on the piston, causing the wheels to turn and the car to move.

Nuclear energy is the energy released from the nucleus of an atom. This is a comparatively new form of energy for us, but it has existed since the beginning of time. Stars shine by the use of nuclear energy.

CONSERVATION OF ENERGY

The conservation of energy states that energy <u>cannot</u> be created or destroyed; however, it can be converted from one form to another. Just as one can take money from a savings account and deposit it into a checking account, energy can be withdrawn from one reservoir and placed into another. It is important to remember that when this is done, no energy is ever gained or lost. <u>You cannot create or destroy energy.</u> This is the <u>first law of thermodynamics.</u>

Imagine a rock sitting on top of a tall ledge. Due to its position, it has potential energy. Once the rock is allowed to fall over the ledge and nears the ground, it loses its potential energy. However, this loss is compensated by a gain in kinetic energy. As the rock nears the ground, it moves faster and faster, thus having a greater and greater amount of kinetic energy.

At the time of impact with the ground, the kinetic energy of the rock will equal the potential energy the rock had before it started to fall. If the rock hits a wood stake, the stake would be forced into the ground. The rock would do work and the kinetic energy is spent. Here, though, is where the analogy of energy being similar to money ends. Once money is spent, it is gone; but once you spend energy to do work, you still have the energy since energy cannot be created or destroyed. But where is it?

A super microscope would show that the atoms which made up the rock, the stake, even the ground nearby are moving faster after the impact. Even though an object as a whole is not moving, the atoms can be vibrating violently. This means that each atom has kinetic energy, the energy of motion.

If one were to total the additional kinetic energy imparted to all the atoms of the rock, stake, ground, and air, it would add up to the exact amount of potential energy the rock had at the beginning when it was still on the ledge. The energy was converted into random kinetic energy -- random motion of atoms.

We do not need a super microscope to detect faster moving atoms. We, as human beings, have special sensors in our skin to detect this movement. We feel moving atoms by the sensation of heat. The faster the atoms are moving, the hotter it feels. When you burn yourself on a hot iron, it is because the atoms are moving so fast that they hurt your skin. <u>Random kinetic energy is heat.</u>

The energy of the rock being converted into heat was not unique. The final form of <u>all</u> energy is random kinetic energy. After a car has been driven and returned to the garage, the chemical potential energy of the gasoline it consumed has been converted into heat. The atoms of the road, the tires, the engine, even the air around the car were made to move faster.

Every day our universe is converting energy into heat. There is more random motion of atoms today than yesterday, and there will be more tomorrow.

HEAT AND TEMPERATURE

Although often confused, heat and temperature are not the same thing. Heat is the total random kinetic energy found in all the atoms in an object. In order to determine the quantity of heat in any object, one could use a giant adding machine to sum the kinetic energy of every single atom. This would be a very tedious process and one that is unnecessary because there are easier methods to estimate the total random kinetic energy.

Temperature is a function of the kinetic energy of each atom. If the atom is oscillating violently, its temperature is high; if the atom is oscillating slowly, its temperature is low.

To help clarify the difference between heat and temperature, consider a burning candle and a swimming pool. Everyone knows the flame in the candle is hot. It has a high temperature, but it does not contain much heat. You certainly could not heat a whole room with one candle! The atoms in the flame (a flame is glowing smoke) are moving around rapidly so the temperature is high. But since there are not many atoms, the total kinetic energy which is heat is low.

In the case of the swimming pool, it does not contain fast moving atoms which means it is not hot, but there is a lot of heat in the pool. Even though each atom does not contain much kinetic energy, when the many, many atoms are summed, the total random kinetic energy (heat) is large.

There are several ways to determine the temperature of an object without really looking at how fast the atoms are vibrating. One method is based on the principle that the faster the atoms vibrate, the more space they require. The atoms do not really increase in size, they just occupy more space because they need more room to oscillate.

This means that when the temperature of an object is increased, it will actually appear larger as each atom moves apart from its neighbor. The hotter things become, the more they expand. The converse is also true; the cooler an object becomes, the more it contracts. There are no exceptions to this, although a few substances such as water at first seem to violate the rule. Water will suddenly become larger when it freezes. This is not due to a violation of the rule that objects contract when cooled, but rather a mechanism involving the spreading apart of the two hydrogen atoms found in each water molecule.

Since things expand when heated, a very simple temperature measuring device can be constructed by placing a liquid such as alcohol or mercury in a hollow tube of glass. As the temperature is increased, the liquid expands so that it rises in the tube; when the temperature is decreased, the liquid contracts and falls down the tube. This tube or thermometer does not measure heat -- it measures temperature.

Although most thermometers are identically constructed, they are marked in different numbering systems. A thermometer that has the number 0 beside the level of liquid when dipped into a pan of ice water, and 100 when dipped into a pan of boiling water at sea level is the Celsius thermometer. If 32 is marked beside the level of the liquid when the thermometer is in the ice water and 212 in the boiling water, it is a Fahrenheit thermometer.

To convert from Celsius to Fahrenheit use:

F = 9/5C + 32

To convert from Fahrenheit to Celsius use:

C = 5/9 (F - 32)

Not all thermometers use the expansion and contraction principle to measure the speed of atoms. One works like a generator. The hotter it becomes, the more electricity it produces. This is called a thermocouple.

Another uses a new material -- liquid crystal. Liquid crystals change color as they are warmed or cooled. Thus by observing these colors, temperatures can be determined.

ABSOLUTE ZERO AND UNITS OF HEAT

What would happen if all the heat were removed from an object? Since heat is the combined random kinetic energy of all the atoms making up an object, the absence of heat would mean that there would be no random kinetic energy. Not a single atom would be moving. Because these atoms would be motionless, it would be impossible for them to move any slower. Therefore, the object would be at the coldest possible temperature. This is referred to as <u>absolute zero</u>, and although it has never been attained, it has been approached within a fraction of a degree.

On the Fahrenheit scale absolute zero is -459.69 degrees; on the Celsius scale it is -273.16 degrees. A third temperature scale, the Kelvin scale, begins at absolute zero. It is impossible to have negative numbers on this scale. Readings of -459.69 degrees Fahrenheit, -273.16 degrees Celsius, and 0 degrees Kelvin all represent the same temperature. This is the coldest possible temperature and it is where atoms cease to move.

When an object is cooled to a temperature approaching absolute zero, some strange effects occur. For instance, a piece of metal can lose its resistance to the flow of electricity and become a <u>super conductor</u>. An electrical current would continuously travel around a loop of wire forming a super conductive magnet. There is currently a great deal of excitement in the physics community because it appears that materials can be turned into super conductors at room temperatures as well as at very cold temperatures. Another result of very cold temperatures is that objects behave as waves allowing them to pass through walls as radio waves can penetrate.

Knowing the definition of temperature and the means by which it can be determined, we now consider the units by which heat is measured. Heat is energy -- the <u>total</u> random kinetic energy contributed by all the atoms in a substance. It could, therefore, be measured in the same units by which any other form of energy is measured -- the foot pound, the joule, and the erg. However, we generally use special units reserved for heat. These are the British Thermal Unit (BTU), the large calorie, and the small calorie.

BTU: The amount of heat required to raise one pound of water one degree Fahrenheit

Large Calorie: The amount of heat required to raise one kilogram of water one degree Celsius

Small Calorie: The amount of heat required to raise one gram of water one degree Celsius

The large calorie is often used when measuring food consumption. Our bodies chemically combine food with oxygen to produce heat. Every large calorie of food will produce enough heat to raise one kilogram of water one degree Celsius when it is used by our metabolism.

When defining the calorie or BTU, the heated substance is always water. Other substances can be heated, but the amount of heat required to raise one pound of another material one degree Fahrenheit could be more or less than a BTU. Likewise, raising the temperature of one kilogram of another material one degree Celsius might require more or less than one large calorie.

The amount of heat necessary to raise the temperature of a certain amount of material is <u>specific heat</u>. A substance with a high specific heat requires more than one calorie or BTU to raise its temperature, while a material with a low specific heat needs only a fraction of a calorie or BTU to increase its temperature. Nearly all materials have a specific heat less than water.

The relationship between regular units of energy and the special ones for heat is called the <u>mechanical equivalent of heat</u>. One small calorie of heat is the same as 4.185 joules of energy. Since there are 1000 small calories in a large calorie, 4185 joules corresponds to one large calorie.

The mechanical equivalent of heat allows us to quantitatively calculate how much heat is produced when we spend energy to do work. Every 4.185 joules of energy will add one small calorie toward the heat death of the universe.

The word <u>entropy</u> is used to describe disorder or chaos. High entropy means disorganization while low entropy means organization. Heat is random disordered motion of atoms so one could refer to the heat death of the universe as the entropy death of the universe.

TRANSFER OF HEAT

The second law of thermodynamics tells us that heat will flow from a higher temperature to a lower temperature when left to nature. Even though heat is present in the air on a very cold day, for instance -20 degrees Fahrenheit, it does not flow into our homes. On its own, heat will flow only from the warm house to the cold air, not from the cold air to the warm house.

It is possible to reverse the flow of heat, but only if outside energy is added. If the cold gas is compressed with a piston in a cylinder, work would be required to force the piston forward to squeeze on the gas. Once this is accomplished, the compressed gas would be warmer. Compressing any gas concentrates the heat energy into faster moving atoms. The compressed warm gas, which was previously cold, could now be used to warm your house. This is the basic principle of the heat pump.

The air conditioner and refrigerator work on the reverse of this principle. If a compressed gas is allowed to expand, the temperature is suddenly reduced. It is important to remember that outside energy is necessary to make a heat pump or refrigerator operate. Usually electrical energy runs the motor which drives the compressor. But remember, without outside energy, heat flows only from hot to cold.

There are three avenues by which heat can flow. The first is conduction, which is similar to the domino effect. For example, if one end of a rod is placed over a candle flame, the fast moving atoms in this end will jar their neighboring atoms. These neighboring atoms in turn jar the atoms next to them that are farther away from the candle. This process continues until eventually the atoms at the other end of the rod start moving faster.

Some materials carry this domino effect faster than others. They are referred to as good conductors of heat. The precious metals such as gold and silver are some of the very best heat conductors. Copper, brass, and aluminum also carry heat well. Generally, if a material is a good conductor of electricity, it is also a good conductor of heat.

If a substance is very poor at conducting heat, it is an insulator. Some examples would be air, fiberglass, and rubber. Using an insulator is an efficient method of minimizing heat loss due to conduction; however, a vacuum is necessary to completely stop the loss by conduction. Placing a vacuum between a hot object and a cold object will stop the domino effect. This can be demonstrated by visualizing a long line of dominoes with a section removed from the middle. When the falling dominoes reach the void, they will stop. The same is true with heat flow by conduction. When the chain reaction of atoms jarring other atoms reaches the vacuum, there are no atoms to jar and therefore conduction would be halted.

Convection, the second means by which heat can flow, is really a special case of conduction. Convection involves moving a liquid or gas over a material to be warmed or cooled. Blowing air over a hot spoonful of soup is an example of heat flow by convection. The heat in the soup flows into the cooler air.

Most homes are heated by convection as hot air or water is forced through the house. Automobiles are usually cooled by this method also. Cool water is circulated around and through the hot engine.

The final process by which heat can flow is <u>radiation</u>. This radiation should not be confused with the radiation from a nuclear reaction. In the case of heat flow, radiation refers to light waves. Not all light waves are detected by our eyes; thus, a warm material could be losing heat by radiation and we would not be able to see the resulting light waves.

When an atom emits a light wave, the atom immediately slows down resulting in its lower temperature. This wave can strike another atom some distance away causing the second atom to move faster and have a higher temperature. Heat was transferred from the first atom to the second by the means of a light wave. This is heat flow by radiation.

The Earth receives its heat from the sun by radiation. The 93-million-mile vacuum between the sun and the Earth prevents heat by conduction or convection but not radiation.

Although conduction and convection can be stopped by a vacuum, radiation cannot because light travels through empty space. Reflective materials can be used to reduce heat loss by radiation. By placing a shiny material around a hot cup of coffee, the material will reflect many of the light waves back into the coffee, thus keeping it warmer. This is the reason why one side of aluminum foil is shiny and also why a thermos bottle has a reflective coating.

PHASES OF MATTER

The speed at which atoms vibrate within a material determine whether the object is a solid, liquid, or gas. Solids contain atoms that are oscillating back and forth; however, as they do this, they remain in one basic area. They are not free to roam within the object. We could visualize the atoms as having invisible springs called bonds which hold them together. These bonds keep the atoms from completely separating, but at the same time, the atoms are free to move back and forth in a confined area. The fact that the atoms are bound together causes a solid to retain its shape.

If the temperature of a solid is increased, the atoms vibrate faster. If they vibrate fast enough, they can break the bonds among them and separate, thus the atoms would fall apart and roll around on top of each other. This is similar to the way a pile of marbles would fall apart if the container holding them were broken. When the atoms of a material start to roll around on top of each other, the material becomes a liquid. To change a solid into a liquid requires the addition of heat in order to raise the temperature. Because the atoms are then free to move around, a liquid will always assume the shape of its container and if there is no container, the atoms spread apart forming a puddle. If more heat is added to a liquid, the atoms can be made to move so violently that they literally fly apart. They jump into the air and form a cloud. This is the gas phase.

The amount of pressure on a substance can affect whether the object is a solid, liquid, or gas; however, temperature is the main variable. If the temperature is raised, the object is changed from a solid to a liquid, the process known as melting. When the temperature is raised further, evaporation occurs, changing the object from a liquid to a gas. The reverse is also true. If the temperature of a gas is reduced condensation occurs, changing the gas into a liquid. A further drop in temperature will freeze the liquid into a solid.

There are a few materials that will bypass the liquid phase, going directly from a solid to a gas when the temperature is increased. This process is sublimation, best exemplified by dry ice.

Liquid, solid, and gas are the three phases of matter which depend on temperature. There are also two other phases or states -- the crushed state and the plasma state.

The crushed state is found inside very massive stars that have collapsed due to their own gravity. The matter is squeezed so tightly that elementary particles such as neutrons, protons, and electrons are compressed into one lump. If the planet Earth were squeezed this tightly, it could be placed into a salt shaker! Because the gravity is so strong from these collapsed massive stars, any light they create is pulled back onto the surface of the star. Because they cannot radiate light into space, they are nicknamed "black holes."

The plasma phase consists of fast moving atoms that have lost one or more electrons. A cloud of plasma gas, sometimes called ionized gas, can be used as a source of thrust in a small rocket engine. Because the plasma phase is charged, it can easily be made to move by placing an opposite charge near by.

PRESSURE

When dealing with liquids and gases, one often deals with <u>pressure</u>. Pressure is sometimes incorrectly used as a synonym for force. One might say that he was forced into doing something, or he might say that he was pressured into doing it. Technically, pressure and force are slightly different. Pressure is defined as the force placed on an object divided by the area over which the force is applied. If the air in a tire exerts 20 pounds of force on every square inch of area inside the tire, we say the pressure is 20 pounds per square inch, abbreviated as 20 psi.

Air does not weigh very much; however, if we consider that our atmosphere extends for many miles into space, the weight of the column of air on top on one's head is tremendous -- over 500 pounds. We could think of this as carrying an invisible brick weighing over 500 pounds on our heads at this very moment. And to make things worse, the air doesn't just push down on us. Because air is a gas, it pushes on all sides as well.

The weight of the air is converted into a pressure by taking the weight of the air and dividing it by the area over which it is applied. This value is around 15 pounds per square inch (14.7 pounds per square inch to be exact) at sea level. If you move above sea level, there is less air on top of you so the pressure is less. Also, weather influences air pressure. People who forecast the weather measure the air pressure using a <u>barometer</u>. When the barometer is high, we usually experience good weather, while a low barometer reading usually indicates stormy weather.

Air pressure is necessary for a soda straw to function. When you use a straw, you first suck in air before any liquid. By removing the air in the straw, the air pressure on the outside of the straw pushes down on the liquid, causing it to be forced up the straw and into your mouth.

When an additional pressure is placed on a liquid or gas in a closed container, the pressure increase will be the same everywhere inside the container. This is <u>Pascal's Law</u>. It can be demonstrated by filling a wood barrel with water and placing a lid with a cork in it on top of the barrel. If one pushes on the cork with 10 pounds of force and if the area of the bottom of the cork is two square inches, an additional pressure of five pounds per square inch is applied to the water. This same additional pressure will be found all along the sides, the bottom, and even along the lid of the barrel. Because the inside of a barrel is thousands of square inches in area, and each square inch has an additional five pounds of force on it, the total force adds up to several tons. The barrel will burst. Pascal's Law is the principle behind any machine that uses hydraulics.

NATURE OF LIQUIDS AND GASES

One characteristic of liquids and gases is their ability to exert a net upward force on objects, causing them in some cases to float. This upward force, called <u>buoyant force</u>, is an indirect result of gravity. When an object is placed in a liquid or gas, the pressure on the bottom is greater than the pressure on the top. This is true because the bottom of the object is at a greater depth than the top; and the deeper the object is submerged, the greater the pressure.

<u>Archimedes</u> (287-212 B.C.) summed this idea in his principle which states that there is a net upward force on an object placed in a liquid or gas. The upward force equals the weight of the displaced liquid or gas. Notice that Archimedes does not concern himself as to whether the object sinks or floats. In either case, there is still an upward force on the object.

If the object floats, then the buoyant force is great enough to cancel out the force of gravity. If the object does <u>not</u> float, it will sink all the way to the bottom, but it will seem to weigh less under the water. One can lift large rocks under water because the buoyant force is present; however, if we attempt to lift the same rock out of the water, the rock loses its buoyant force and we experience its full weight.

Archimedes tells us that the amount of buoyant force depends on the weight of the displaced material. Two objects cannot occupy the same space at the same time; therefore, if an object is placed in water, the water is pushed aside to make room for the object. This water rises up the sides of an open container, and if the sides are not high enough, the water spills over the top. The weight of the water that is displaced equals the buoyant force pushing upward on the object.

For example, if a stone weighs 25 pounds out of water and seems to weigh only 15 pounds in water, then 10 pounds of water was displaced. If another stone weighs 40 pounds out of water and displaces 32 pounds of water, it would seem to weigh only 8 pounds when submerged.

Since the buoyant force is a function of the weight of the displaced material, increasing the weight of the displaced material would increase the buoyant force. If salt is added to water, it makes the water heavier. Thus when an object is placed in salt water, the buoyant force will be greater than when the same object is placed in pure water. This is why it is easier to stay afloat in salt water. In a very heavy liquid such as mercury, bricks actually float.

Another useful law concerning liquids and gases is <u>Bernoulli's Theorem</u> (1700-1782). It states that whenever a liquid or gas moves parallel to a surface, there will be a reduction in the pressure on the surface. A piece of paper suspended horizontally has air pressure pushing down on the top as well as air pressure on the bottom pushing upward. These two pressures cancel each other. If by some means, one of these air pressures could be reduced or eliminated, then the pressure which remains on the other side would cause the paper to move. By merely blowing across the top of the paper, the air pressure on the top is blown away and the remaining pressure on the bottom causes the paper to rise. Likewise, blowing across the bottom of the paper would reduce the pressure on the bottom, causing the paper to be pushed downward.

While a car is moving, air rushes by both sides. According to Bernoulli's Theorem, this means that there is a reduction of pressure on both sides of the automobile. This is the reason why a piece of paper is sucked out of an open window of a moving car.

Perhaps the most practical application of Bernoulli's Theorem is the airplane. See Diagram 24A.

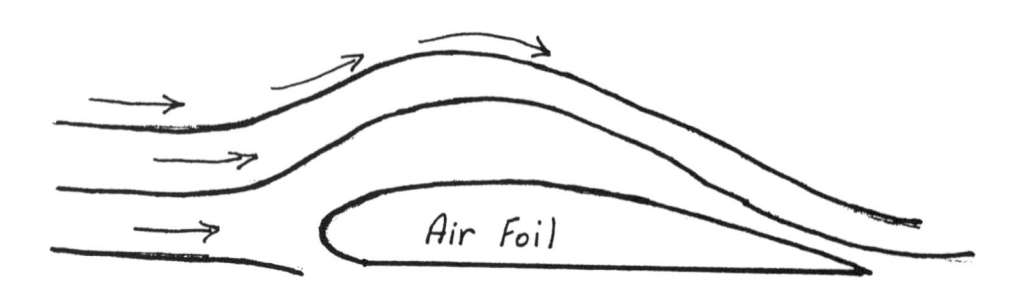

DIAGRAM 24-A

The design of an airplane's wing is such that the bottom is flat and the top is curved. This shape is called the _air foil_. As the wing moves forward, air rushes over the curved top of the wing faster than the air moves over the flat bottom. This faster moving air on top reduces the pressure significantly more than the slow moving air at the bottom. Thus, there is more pressure remaining on the bottom and this gives the airplane lift.

INTRODUCTION TO WAVES -- PART I [START]

A wave simply defined is a disturbance. When one drops a rock into a pond, disturbances emulate from the point of entrance generating ripples that move across the surface. If you shake one end of a stretched rope, disturbances can be seen moving toward the other end.

When a wave is propagating, the material that is being disturbed is the <u>medium</u>. Water is the medium for water waves, a rope is the medium for rope waves, and air is usually the medium for sound waves.

When the medium vibrates it usually oscillates in a fashion similar to water waves. The medium vibrates up into the shape of a mountain and then down into a valley. The mountain is the <u>crest</u> of the wave and the valley is the <u>trough</u>. The two together make one <u>cycle</u>. The length of the cycle is the <u>wavelength</u> and the height is the <u>amplitude</u>. See Diagram 25A.

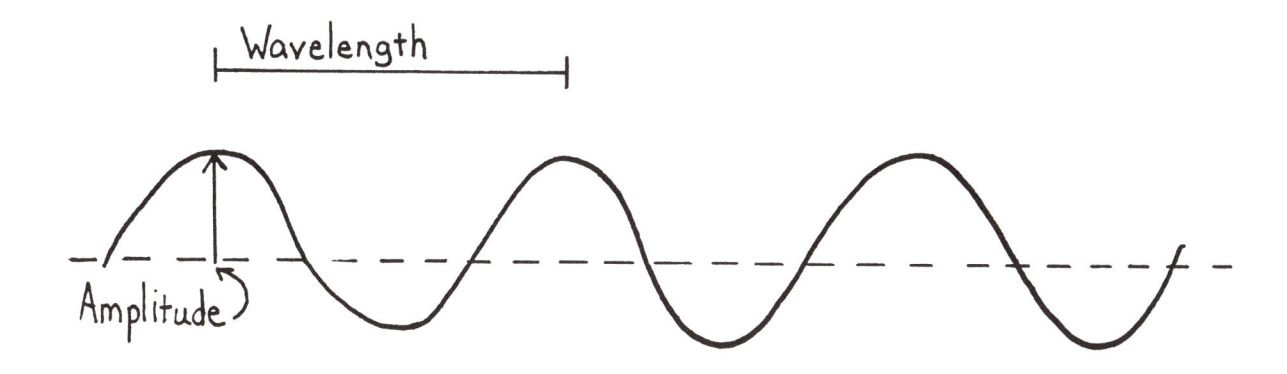

DIAGRAM 25-A

The time it takes the medium to vibrate through one cycle is the <u>wave period</u> and the number of cycles which pass a point in a given length of time is the <u>frequency</u> of the wave. For example, if 20 cycles pass in front of you in one second, the frequency would be 20 cycles per second. A relatively new unit in physics is the <u>Hertz</u>, abbreviated Hz. It means cycles per second; therefore, 20 cycles per second could be stated as 20 Hertz or abbreviated 20 Hz.

It is important to keep in mind that the medium does not undergo any net movement. As a wave passes through a medium, it merely vibrates it. When a wave moves across a swimming pool, water is <u>not</u> transported from one side to the other; instead, it is the wave that moves from one point in the water to the next, only giving the water the appearance of movement.

The motion of a wave is similar to the appearance of a moving arrow in a large array of light bulbs. By turning the lights on in a specified sequence, the arrow appears to move across the array. As this occurs, none of the light bulbs are actually moving, it is only that the lighting passes from one bulb to the next. This is true for a wave. The medium does not undergo any net movement; the disturbance just passes from one point to the next.

A log in the water only vibrates up and down with the waves as they move by. The log is not pushed forward. This seems contrary to what is observed when a surfer is pushed into the shore by an ocean wave. The contradiction is explained when one realizes that ocean waves near the shore are not just waves, but water currents as well. When the medium is actually moving, we call it a current. Water moving in a stream is an example of a current of water. Waves in the middle of the ocean are strictly waves. They do not push forward and thus are useless for surfing. As the waves near the shore, the bottom of the waves are affected by the shoreline. They tip at the top which is known as breaking. This produces moving water and thus a water current that can push forward on the surfboard.

When the medium vibrates at right angles to the direction the wave is moving, the wave is a transverse wave. Waves in a rope are examples of transverse waves.

If the medium vibrates back and forth parallel to the direction the wave is moving, it is a longitudinal wave. If a stretched slinky has one end pushed forward and backward, this will start a wave motion where the coils next to the end will vibrate forward and backward. Then the next coils will move forward and backward, and then the next, and so on. The wave moves parallel to the direction the coils are oscillating. Waves in the air which our ears interpret as sound are a good example of longitudinal waves. When you speak, your vocal cords vibrate the air back and forth several hundred times a second. This vibration moves through the air until the air that is next to a listener's ear is made to vibrate back and forth.

Waves travel through a medium at finite speeds. Water waves pass through water at only a few miles per hour. Air waves travel through air about 700 miles per hour or about 1100 feet per second, although this varies with the density of the air.

If an object travels faster than the waves can propagate, the waves in front of the object are bunched together to create a large amplitude wave known as a shock wave. We see this when a boat moves faster than a few miles per hour. The shock waves move back along the side of the boat and can be easily seen. An airplane traveling faster than sound will produce a shock wave in the air which is known as a sonic boom.

INTRODUCTION TO WAVES -- PART II

Sound waves and light waves are extremely important to us because much of what we know comes to us by sound or light. We shall see later that light is a transverse wave that is doubly unique because of its great speed and its ability to pass through a vacuum.

Sound travels at a much slower speed than light. Its exact speed depends on the medium. In air, sound travels about 1100 feet per second or approximately one mile in five seconds. Sound can travel in any liquid, solid, or gas; however, it cannot pass through a vacuum. Sound is a longitudinal wave which means as sound waves approach your ears, the air is vibrating forward and backward, not up and down.

How we hear a sound wave depends on its frequency. If the air is vibrating back and forth less than about 20 times a second, our ears do not pick up the sound -- we hear nothing. If the frequency is increased above 20 Hertz we can hear it, but the sound is low in pitch. As the frequency is further increased, the pitch also increases. When a frequency of about 20,000 Hertz is reached, we are no longer capable of hearing the sound. These high-pitched sounds which we cannot hear are called <u>ultrasonic sound</u> waves.

Some animals such as the dog and most insects can hear ultrasonic sound waves. This is the principle of the dog whistle. It generates a sound wave whose frequency is about 22,000 Hertz. We cannot hear it, but the dog can. Ultrasonic waves can be used under water as radar is used on land or in the air. This is the principle of sonar. Jewelers use ultrasonic waves for cleaning, and in medicine they are used in a manner similar to X-rays.

When an object is struck it will oscillate back and forth. The frequency of this oscillation is the <u>resonant frequency</u> of the object, which is its own natural frequency of oscillation and every object has its own resonant frequency. A glass half filled with water will vibrate at a different frequency than a glass that is empty. This is why music can be made by striking various glasses with different amounts of water. The more mass in an object, the lower its resonant frequency, thus the more water in the glass, the lower the resonant frequency.

If a sound wave happens to have the same frequency as the resonant frequency of an object, the wave can cause the object to shake back and forth and even break. This is the principle used by opera singers to break a champagne glass. They must produce a note whose frequency matches the natural frequency of the glass.

ELECTROMAGNETIC WAVES

The <u>electromagnetic wave</u> (or EM wave, for short) is a double transverse wave which travels through a vacuum at 186,282 miles per second which is usually rounded off to 186,000 miles per second. It is the fastest thing known and the only wave that can pass through empty space.

Because a wave is defined as a disturbance, a perplexing question arises when an electromagnetic wave moves through a vacuum -- what is being disturbed? If a vacuum is empty space, what are we shaking or vibrating? During the nineteenth century scientists answered the question by postulating the existence of the <u>ether</u>. Ether was defined as an invisible and massless entity which filled the voids of space. An electromagnetic wave was considered a disturbance that propagated through the ether.

In the second half of the nineteenth century James Clerk Maxwell (1831-1879) introduced ideas that could explain the movement of a light wave without the ether. He interpreted an electromagnetic wave as an oscillating force field rather than a disturbance of some medium. It is known that a negative particle such as an electron exerts a force of repulsion on other negative particles and a force of attraction on positively charged particles (like charges repel; opposites attract). The effect of this force (force field) radiates outward from the electron at the speed of 186,000 miles per second. If two electrons were placed 186,000 miles apart, it would take one second for them to start repelling one another. Now if an electron were moved back and forth, the force field would also oscillate, producing what could be called an electric wave.

Maxwell was able to show that this oscillating electric wave would produce a force field similar to the force field around a magnet. Therefore, when a charged particle is made to vibrate, it produces two oscillating force fields -- the electric and the magnetic. Maxwell assumed this was an electromagnetic wave. See Diagram 27A.

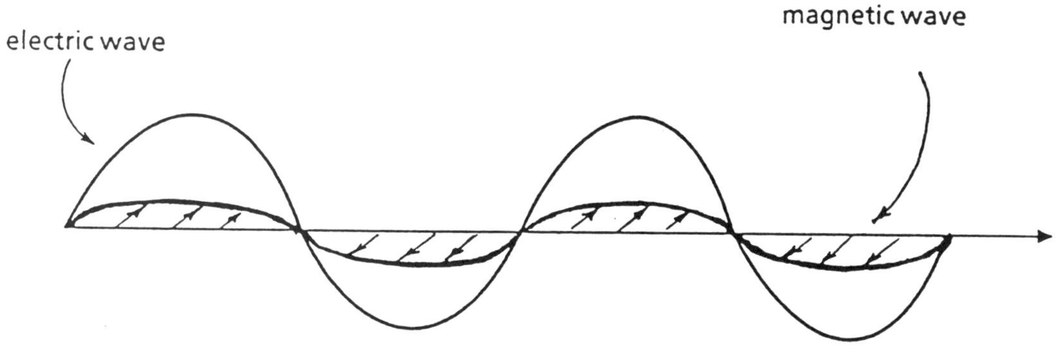

DIAGRAM 27-A

Although a few scientists may still like the idea of vibrating ether as the electromagnetic wave, nearly all believe that an EM wave is an oscillating electric and magnetic force field. And since force fields can travel through a vacuum, there is no longer any need for the ether.

As humans we have special sensors which detect some, but not all electromagnetic waves. The sensors are our eyes and the electromagnetic waves we can see are the visible light waves. Even though the electromagnetic wave may be vibrating force fields, we still use terms such as wavelength, amplitude, wave period, and frequency. Just as our ears can tell the frequency of a sound wave by pitch, our eyes can tell the frequency of a visible light wave by its color. If the frequency is about 400 trillion Hertz we see red, 450 trillion Hertz produces orange, 500 trillion Hertz yellow, and 550 trillion Hertz is seen as green. Blue is next at about 600 trillion Hertz, then indigo at 650 trillion Hertz. Finally violet is the highest frequency that we can see. It is about 700 trillion Hertz. To help remember the visible light waves in order from lowest frequency to highest, remember the name ROY G BIV. The R is for red, O is for orange, Y is for yellow, and so on.

Black and white are not colors. Black is what we perceive when there is no visible light and white is what our eyes perceive when they are saturated with all colors. We can trick our eyes into seeing all colors and thus producing white just by looking at a combination of red, green, and blue. These are the three primary lights which should not be confused with the primary pigments which artists use -- cyan, yellow, and magenta.

THE ELECTROMAGNETIC SPECTRUM

The human eye detects only a very small portion of existing electromagnetic waves -- only those which vibrate between 400 trillion and 750 trillion times a second. If we could see <u>all</u> electromagnetic waves for just one minute, our entire concept of the physical universe would be drastically changed. We wake up in the morning and worry about how we look in terms of visible light waves, but what about the other electromagnetic waves? A person might be handsome or beautiful when viewed by regular light, but might appear completely different when viewed by other EM waves.

A list containing all electromagnetic waves in order from low frequency, which corresponds to long wavelength, to high frequency with short wavelength is the <u>electromagnetic spectrum</u>. The spectrum begins with <u>radio waves</u> which are used for communication. A radio wave is just like a visible light wave, but it is too low in frequency for us to see. Generally they range from zero to millions of cycles per second (M Hz). Radio waves can carry information by changing the amplitude (size of the wave) or by changing the frequency of the wave. Those stations that broadcast by changing the size are AM (amplitude modulation) stations and those that change the frequency are FM (frequency modulation) stations. We can listen to an AM or FM radio station inside a building because radio waves can penetrate most non-metallic materials.

If we increase the frequency to several billions of cycles per second, we have the <u>microwave</u> and if we go into trillions of Hz, we have the <u>heat wave</u>. Heat waves are not hot; they just produce a lot of heat when they strike something. All EM waves can jar atoms and thus produce heat, but usually at a much lower magnitude than the heat wave.

Notice that the microwave is sandwiched in between the radio wave and heat wave; thus having some of the properties of both. This is why the microwave is excellent for cooking. It can penetrate and heat all layers of our food simultaneously. A standard oven heats only the surface. The heat then works its way slowly to the inner layers by conduction.

When the frequency is increased to hundreds of trillions of cycles per second, but still less than the 400 trillion Hz threshold of the human eye, it is an <u>infrared</u> wave. This wave usually heats the surface of objects and is used in fast food restaurants to keep food warm. The sun emits large quantities of infrared waves that can produce sun burns.

Between 400 trillion and 750 trillion Hz, we have the visible light waves; however once we increase the frequency of an EM wave beyond 750 trillion Hz, it becomes invisible again. This is the <u>ultraviolet</u> wave which is often incorrectly called "black light." Small doses of ultraviolet light are helpful for plant life as well as animal life. It is responsible for producing a suntan; however, large quantities can destroy cells and thus the ultraviolet light is sometimes used to sterilize objects.

Beyond the ultraviolet is the <u>X-ray</u>. When X-rays were discovered in 1895 by the German physicist Wilhelm Roentgen (1845-1923), he thought he had discovered some new mysterious ray. Instead the X-ray proved to be nothing more than a very high frequency EM wave. The German physicist Max Karl Planck (1858-1947) stated that the higher the frequency of a wave, the greater its energy. Therefore, X-rays are very energetic and thus dangerous to living tissue. One should not be exposed to these waves any more than absolutely necessary.

At the end of the electromagnetic spectrum is the gamma ray. It too was incorrectly identified as a ray, but also proved to be nothing more than the highest frequency, and therefore most energetic electromagnetic wave. We are all occasionally exposed to gamma rays, as they seem to be present throughout the cosmos. It is possible that a reproductive cell in your great grandmother was struck by a gamma ray, causing the color of your hair to be different from what it otherwise would have been.

COHERENT LIGHT AND LASER

We shall now consider twelve wave occurrences or twelve wave phenomena. The first is <u>luminosity</u> which refers to an object that is radiating an electromagnetic wave. This can be done by one of four ways.

First, the object can be heated. Any object above absolute zero radiates an EM wave. Generally the frequency of the wave is low and thus cannot be seen. Not until the object is heated to hundreds of degrees, does the radiated frequency become detectable by our eyes. This is what happens when an iron rod is placed in a flame. Long before we see it glowing, it produces infrared waves. As it becomes hotter, it begins to send out red light waves. Increasing the temperature further will cause it to turn orange and then yellow. When green and blue light waves are added at a higher temperature, all the colors blend together causing the rod to appear white.

Most light bulbs contain a piece of wire made of tungsten. This tungsten filament is heated as electricity passes through it, causing the light bulb to be luminous. The tungsten filament mainly produces red, orange, yellow, and green light. Because it does not produce as much blue and violet, the light from a tungsten filament seems to be a yellow white, rather than a pure white.

A second means by which luminosity can be produced is to pass charged particles through a gas. Electricity which is a stream of negative electrons can cause almost any gas to glow. Electricity sent through neon gas produces an orange light; electricity through sodium produces yellow; and electricity through hydrogen can produce blue. Charged particles from the sun often rain down through the gases of our atmosphere producing the Northern and Southern Lights.

A third method is a chemical reaction. By combining certain chemicals, light can be produced. We see this in the lightning bug and in some deep-water fish.

Finally, a nuclear reaction can generate an electromagnetic reaction. On Earth we do not encounter this reaction very often, but throughout the entire physical universe this is the most prevalent source of luminosity. The stars, including our sun, shine as a result of reactions that take place in the nuclei of atoms.

No discussion of luminosity would be complete without mentioning a powerful kind of light produced by making all the waves vibrate in phase with one another. Soldiers marching in step produce a great force on the ground as they all pound the earth simultaneously. This force could destroy a small bridge; thus soldiers usually break cadence when crossing such a bridge.

Almost all light consists of waves that are out of step or out of phase. One wave is going through a crest, another is going through a trough, while still another is somewhere in the middle. This is <u>incoherent light</u>. If somehow these light waves could be made to march in step, they would produce a beam of <u>coherent light</u> that could exert a large force on an object.

To produce coherent light many atoms that are just about ready to produce light waves are assembled together. One atom will spontaneously emit one wave, which will pass by a second atom causing the second atom to emit a wave that is exactly in step with the first wave. These two coherent waves will pass by a third atom stimulating it to emit another wave in step with the previous two. This continues until millions of waves are marching together producing a powerful beam. The device used to do this is named for the process -- light amplification by the stimulated emission of radiation, or LASER for short.

LASERS have many useful purposes besides blasting holes through walls. They are presently being used in communication. LASERS that can be turned on and off by our voice are projected through special reflective pipes or fibers that send the light from one city to another. When the modulated laser light is received at the other end of the reflective pipe, it is changed back into voice.

LASERS have medical applications such as burning away tumors and welding detached retinas. They are also used in the entertainment world. The CD (compact disk) player uses a LASER by bouncing a LASER beam off of the CD. As the beam reflects from varied surfaces on the disk, sounds are decoded. Also LASERS are used to make holograms. A hologram yields a true 3D picture that enables depth perception, and gives the ability to see above, below, and behind objects just by viewing them from different angles.

REFLECTION, TRANSMISSION, AND ABSORPTION

The first of the twelve wave occurrences that we studied was luminosity. The second is reflection. When a wave hits an object and returns, we say it has been reflected. A reflected sound wave is an echo and reflected light waves are the means by which we see most objects. Only things such as light bulbs, flames, and other glowing substances produce visible light. All other objects which we see merely reflect visible light.

The wave that hits an object is the incident wave and the wave that bounces off the object is the reflected wave. The angle between the incident wave and the surface will always be equal to the angle between the reflected wave and the surface. See Diagram 30A.

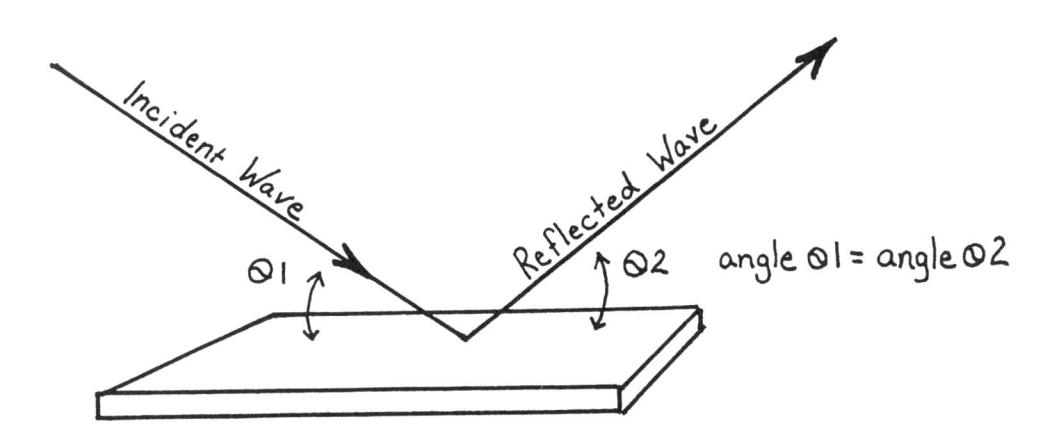

DIAGRAM 30-A

Generally objects reflect less than 100 per cent of the light that shines upon them. Even a shiny mirror does not reflect all of the light that strikes it; however, the Dupont Company has developed a material called fiber optics that virtually reflects all light. A long flexible tube or pipe can be made from this material. Light can be projected into an end and it will reflect back and forth inside until it comes out the other end. Fiber optics are used in medicine, communication, and decorative lamps.

Reflection is often <u>frequency dependent</u>. A material may reflect one frequency but not another. Imagine a piece of paper that reflects a light wave only if its frequency is about 400 trillion Hertz. If we shine white light onto this paper, how will it appear to us? White light is a blend of all colors and thus all frequencies between 400 trillion and 750 trillion Hertz. Since red light is an EM wave oscillating about 400 trillion Hertz, the paper that reflects only 400 trillion Hertz appears red to us. A yellow piece of paper only reflects yellow light (500 trillion Hertz) and a blue piece of paper only reflects blue light (600 trillion Hertz). A white piece of paper reflects all frequencies of visible light while a black piece reflects none.

It is interesting to speculate what we would see if green light were projected onto red paper. In fact, the paper would probably be faintly seen as red since it is very difficult to manufacture a light bulb that emits only one color. The light from the bulb may look green to us, but it probably contains small quantities of other colors; thus, the small amount of red in the green light will be reflected by the paper allowing us to see it as a faint red. If, however, we could obtain a beam of light that is pure green and shine it on paper that reflects only red light, the paper will appear black. This is because the red paper does not reflect green and therefore no light is reflected.

When light is scattered in many different directions by a rough surface such as piece of paper, we call it <u>diffuse reflection</u>. A shiny surface such as a mirror does not scatter the waves and this is why we can see an image of ourselves in a mirror, but not in a piece of paper.

If a wave is not reflected, it may be absorbed. <u>Absorption</u> is the third wave occurrence or phenomena. It happens when a wave knocks the atoms of a material causing an increase in heat. The more light that is absorbed, the hotter the object will become. This is the reason black clothing feels warmer than white clothing when worn out in the sun. The white clothing reflects the light waves while black clothing absorbs the light waves converting their energy into heat.

If a material does not reflect or absorb a wave, there is only one option remaining. The wave may pass through the object. This is <u>transmission</u>, the fourth wave occurrence. Light passing through a glass is an example of this. Transmission is also frequency dependent. A material may act as a selective roadblock or filter. Some frequencies may pass through while others are absorbed or reflected.

PHOSPHORESCENCE, FLUORESCENCE, AND DOPPLER EFFECT

The fifth wave occurrence or wave phenomenon is <u>phosphorescence</u>. This can be defined as delayed reflection. If the molecular structure of a material is just right, light waves can be temporarily captured and held for a period of time after which they are sent back to the outside world. Phosphorescent materials are often seen glowing in the dark. This glowing is the result of light waves that struck the object earlier.

At one time watch dials were painted with phosphorescent paint. One could hold the watch under a bright light, then walk into a dark room and the dial would seem to be glowing, usually a pale green color. Golf balls, tennis balls, baseballs, and Frisbees can be painted with phosphorescent paint and by first holding them under a bright light, they could be used at night.

<u>Fluorescence</u> is the sixth of the wave occurrences. It is defined as a change in frequency during reflection. Certain materials can actually change the frequency of the wave they reflect. If white light is projected onto an orange fluorescent piece of paper, instead of just reflecting the orange light waves as in the case of a piece of ordinary orange paper, the fluorescent orange will in addition, change the red, yellow, green, blue, indigo, and violet that is part of white light into orange. For this reason, the orange seen from a fluorescent orange paper under a white light will be very bright. In the same way, green fluorescent paper will change light to green and blue fluorescent paper will change light to blue.

An interesting as well as artistic use of the fluorescent effect is achieved when ultraviolet light or X-rays are projected onto fluorescent materials. Since ultraviolet light and X-rays are too high in frequency for our eyes to detect, we do not see the incident light; however, we do see the light returning from the material after there has been a change in frequency.

There is another way to change the frequency of waves in addition to the fluorescent effect. This is the seventh occurrence which is the <u>Doppler Effect</u>, named for the Austrian physicist who first published his work in this area in 1842. The Doppler Effect is defined as a change in frequency due to relative motion.

We will begin with its effect on sound waves and later apply it to electromagnetic waves. If a train is rushing toward you and its whistle is blown, the sound waves in front of it are crowded together producing a shorter wavelength. This in turn causes more cycles of the wave to hit your eardrum in a second, producing a higher frequency. The whistle actually changes its pitch.

If the train were rushing away from you, the cycles would spread apart producing a lower frequency. The sound would have a lower pitch. This change in frequency due to motion is independent of who or what is moving. The same effect is produced whether you move toward or away from the train whistle.

We hear the Doppler Effect when a race car passes us. The engine sounds high in pitch as it approaches us and immediately drops in pitch as it passes.

BIG BANG THEORY AND REFRACTION

The Doppler Effect does not occur just with sound waves. EM waves can also be Doppler shifted. Imagine a light that is emitting green electromagnetic waves. If the light were rushing toward you, the cycles of the green light would be pushed together producing the effect of a higher frequency; thus, you might see the light as blue or violet instead of green. This would also happen if you were rushing toward the light. If the light were moving away from you or if you were moving away from it, the cycles would be spread apart producing a lower frequency. This lower frequency light may appear as yellow, orange, or red.

It must be pointed out that we do not see these changes in color in everyday life because the speeds required to produce a noticeable Doppler shift in light waves are much greater than the speeds required to produce a noticeable Doppler shift in sound waves. However, electronic equipment, can measure the Doppler Effect of light waves produced by objects moving at slow velocities. If you received a ticket for speeding, you were probably caught by a radio wave that was Doppler shifted as it bounced off the front of your car. Some weather radar stations called Doppler radar can detect a tornado by sensing wind currents moving in opposite directions in the base of the tornado.

The Doppler Effect has given us a great clue as to the origin of the physical universe. As we look at distant galaxies, we notice the light is shifted toward a lower frequency or toward the color red. This red shift suggests to us that all of the galaxies were once compressed together into one small very dense ball until a great explosion occurred throwing everything outward. We are still seeing this outward rush as evidenced by the Doppler Effect. The explosion has been nicknamed the Big Bang and is estimated to have taken place 13 to 18 billion years ago.

Refraction, the eighth occurrence, is the process in which light is bent as it travels from one medium to another. When a flashlight is aimed into a glass of water, the water will refract the beam toward the normal line. The normal line is the line perpendicular to the surface. See Diagram 32A.

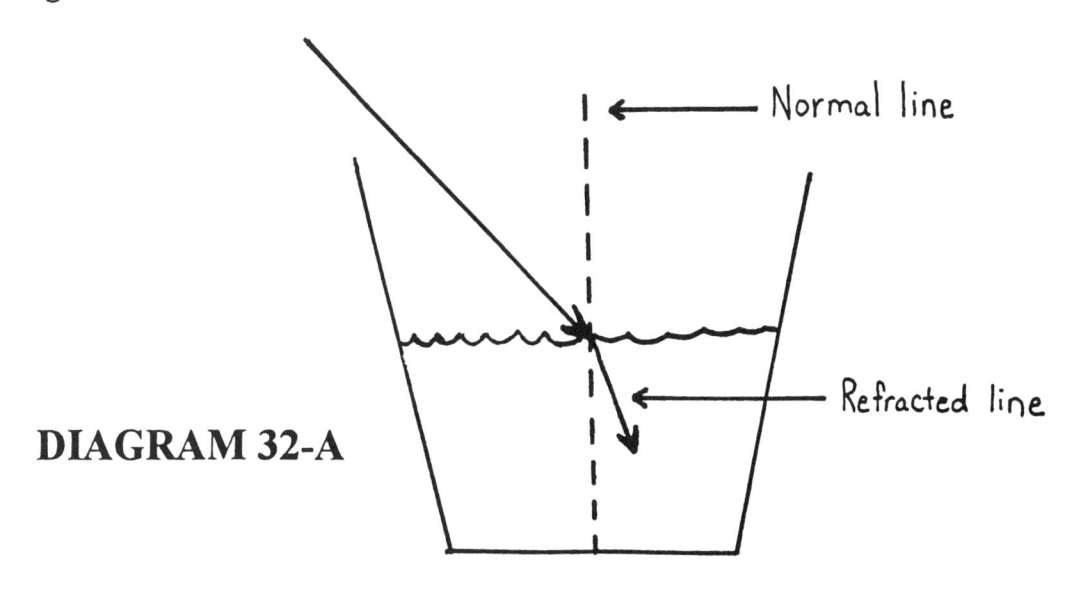

DIAGRAM 32-A

When a wave passes in the opposite direction, that is from the thick water to the thin air, the bent or refracted wave is bent away from the normal line. This happens because a wave passing from low-density material to high-density material is refracted toward the normal, while a wave passing from a high-density material to a low-density material is refracted away from the normal. If a wave is aimed right along the normal and thus enters a new material at an angle of 90 degrees, there is no refraction.

We have stated several times that white light is a composite of all the colors of the rainbow. This can now be demonstrated using refraction. High frequency waves are refracted more than low frequency waves; thus, the low frequency red light in a white beam will be bent less than the high frequency violet. All the colors will be spread out producing a rainbow or spectrum of visible light. This is how a rainbow is produced in nature. The white light from the sun is refracted by falling water droplets. Ice crystals in high altitude clouds can create this same effect, producing a rainbow in the shape of a ring around the sun or moon.

DIFFRACTION, POLARIZATION, AND INTERFERENCE

In addition to refraction, there is another way to bend the path of a wave. This is <u>diffraction</u>, occurrence number nine. Diffraction is the bending of a wave as it passes around the edge of an object. Your voice will bend around the corner of a building just as light will bend around the edge of a piece of paper.

The angle of the bend due to diffraction depends on the frequency of the wave, as was the case with refraction. <u>With refraction the higher the frequency, the greater the bend; however, with diffraction, the lower the frequency, the larger the bend</u>. Low pitch sounds will bend around corners better than high pitch sounds. This is one reason why low pitch sounds can be heard at great distances better than high pitch sounds.

If one puts a very small hole about the size of a pinhead in a piece of paper, one can see how low frequency light waves are bent more than high frequency light waves when diffracted. When white light shines through the hole, the inner edges diffract the light causing the colors to spread out into a rainbow ring. The red is on the outer circumference because it has the lowest frequency and is diffracted the most. The inner circumference is violet since it has the highest frequency. See Diagram 33A.

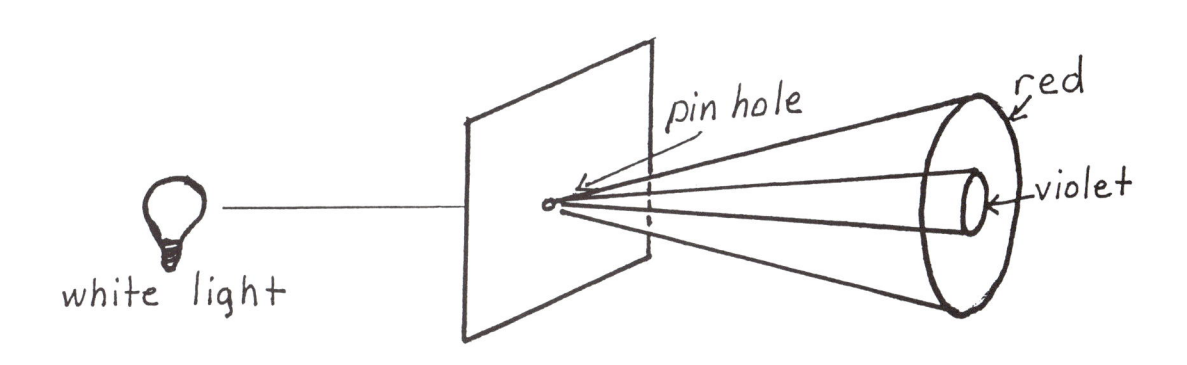

DIAGRAM 33-A

The next wave occurrence, number ten, is <u>polarization</u>. This deals with the direction a wave is vibrating. A rope wave can be made to vibrate up and down, sideways, or even diagonally.

When a large number of waves are vibrating, some with up and down motion, some with sideways motion, and some with diagonal motion, we say the waves are non-polarized. If, however, all the waves vibrate in the same direction or in the same plane, then we say the wave is polarized.

Light is generally non-polarized, but this can be changed. By projecting a beam of non-polarized light through thin slits produced by long chains of hydrocarbon molecules, all the EM waves are absorbed except those waves that have the electric portion vibrating parallel to the slits. The magnetic portion of the electromagnetic wave is much smaller than the electric portion and therefore the magnetic portion is ignored.

Some plastic filters, often referred to as polarization filters, can be constructed so that it contains these slits between hydrocarbon chains. It is interesting to note that if two polarization filters are placed on top of each other, light may or may not pass through them. It depends on the manner in which the filters are placed together. If the slits in both filters are aligned in the same direction, light will make it through; however, if one is placed so that its slits cross over the slits of the other, no light makes it through.

<u>Interference</u> is the eleventh occurrence. When two waves meet, they interfere with each other by producing a new shape. This new shape only lasts for an instant because <u>the two waves will pass through each other after the interference</u>. If two crests come together, the new shape is a large crest; likewise, two troughs will create a large trough. This is often called constructive interference because a larger wave is constructed. The large crest or the large trough only exists for an instant, because the two crests or two troughs will pass through each other. In Diagram 33B we see the results of two crests coming together in a slinky. The big crest is created only at the instant the crests meet. After they meet, the two crests continue on as if the other never existed.

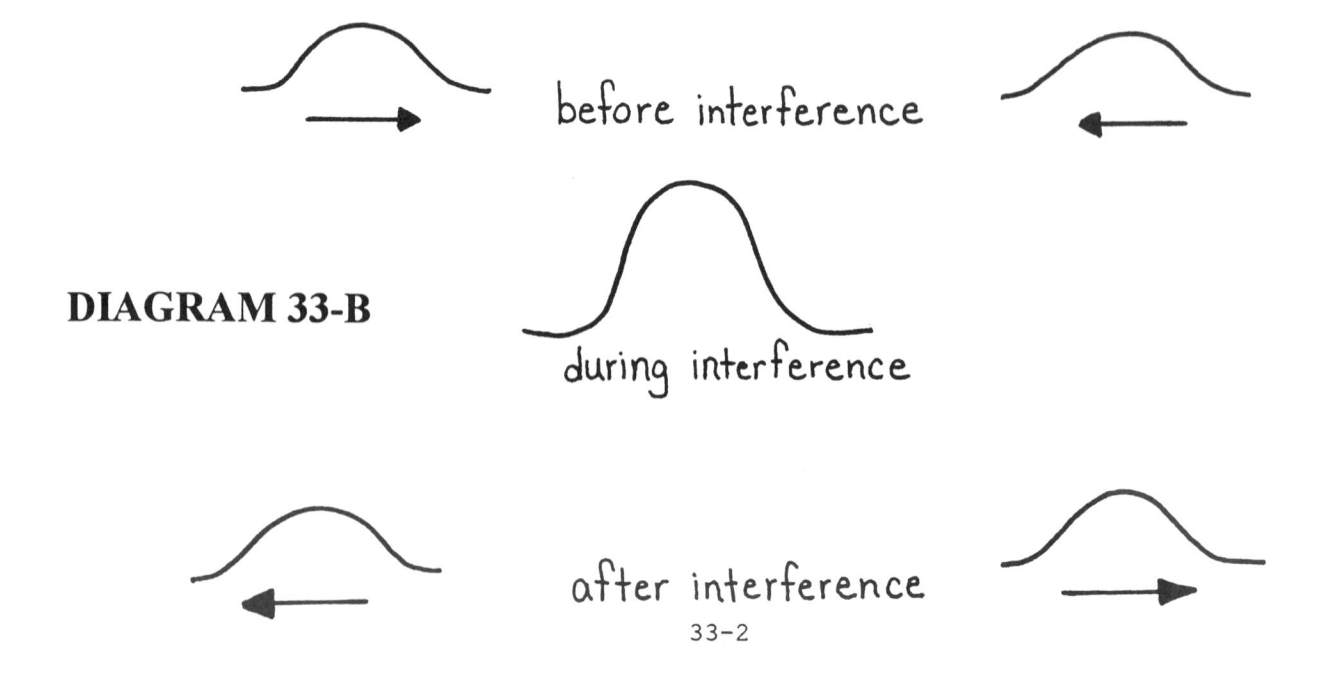

DIAGRAM 33-B

before interference

during interference

after interference

When a crest and a trough of the same size meet, destructive interference occurs. The crest and trough cancel each other, but again only for an instant. They then reappear and move on. This is called destructive interference. See Diagram 33C.

Interference can take place with light waves. Waves that meet constructively produce large waves which appear as bright light. Light waves which meet out of phase or destructively cancel each other out and produce black. The fact that EM waves can interfere with one another is perhaps the most convincing argument for accepting light as a wave.

before interference

during interference

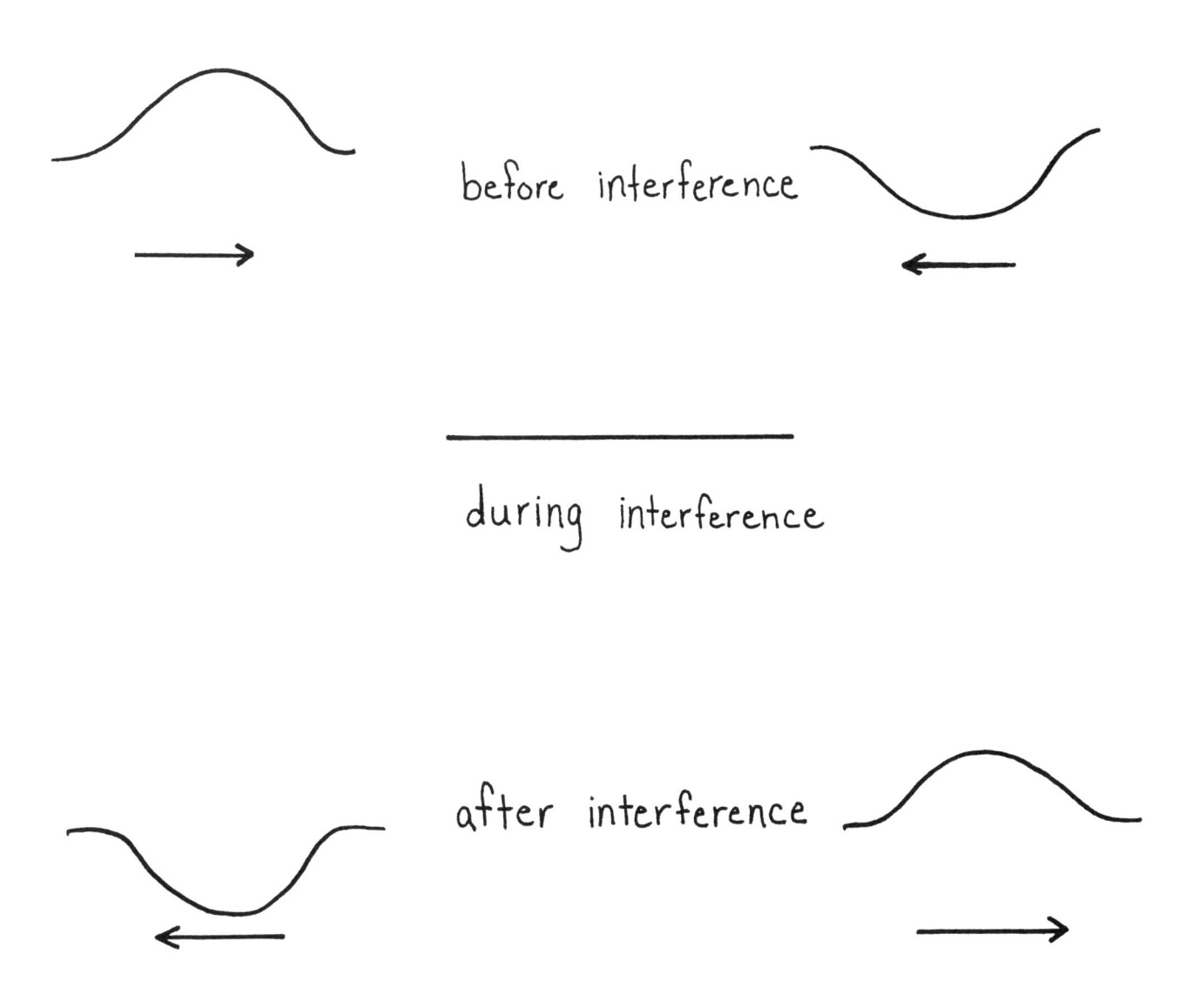

after interference

DIAGRAM 33-C

PHOTOELECTRIC EFFECT AND BOHR THEORY

The twelfth and final wave phenomenon is called the <u>photoelectric effect</u>. This occurrence takes place only with EM waves. Shortly after the turn of the century scientists noticed that a beam of light had the ability to knock an electron out of an atom. Supposedly this is impossible because light is a wave and waves cannot push an object forward; it only causes an object to vibrate.

When it was discovered that light waves are different and that they can push on objects, physicists were startled. At first they thought of dismissing the idea that light is a wave, but they decided against it because they would not be able to explain the diffraction of light or the interference of light. A young scientist at the time had a theory that EM waves are more than just a wave. He suggested that EM waves contain little packages of energy that act like bullets that exert a forward push when light strikes an object. The little bullets are called <u>photons</u> and the idea is now widely accepted. When you aim a flashlight at someone, you are spraying them with photons. The young scientist won the 1921 Nobel Prize in physics along with a great deal of fame. His name was Albert Einstein.

The photoelectric effect is used in solar cells and so-called electric eyes to produce electricity from light. Light waves with photons strike the solar cells ejecting electrons which make up an electrical current. Most satellites produce their electricity by means of solar cells on long extended panels.

Sometimes when a photon strikes an atom the electron is not kicked completely out of the atom, but is moved to an orbit which is further away from the nucleus of the atom. <u>Niels Bohr</u> (1885-1962) suggested that the electrons go around the nucleus in different sized orbits, or energy shells as he called them. The orbit closest to the nucleus is the K shell, then the L, M, N, O, P and Q. See Diagram 34A.

DIAGRAM 34-A

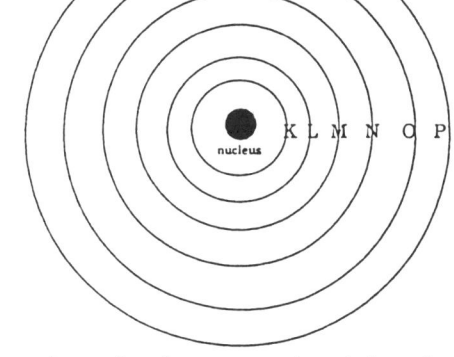

The energy of the incoming photon will determine what happens when it bombards an electron. If the photon is from a low frequency EM wave, the electron might be pushed outward just one shell, that is from K to L. If the photon is from a higher frequency EM wave, the electron may be pushed outward two or three shells. If the photon is from a very high frequency EM wave, the photon is energetic and might kick an electron out of an atom, producing the photoelectric effect.

34-1

THE ULTRAVIOLET CATASTROPHE AND QUANTUM MECHANICS

It is common knowledge that an iron rod placed in a fire will begin to glow. As the temperature of the rod increases, two things happen. First, the frequency of light radiated by the rod increases; and secondly, the brightness of light increases.

When the iron rod becomes just hot enough to glow, it appears as dull red. As it becomes hotter, it turns orange, and this orange light is brighter than the previous red. A further increase in temperature causes the rod to appear yellow, and again it will be much brighter. As the temperature is steadily increased, the predominant color of the rod continues shifting to a higher frequency and the brightness continues to increase in magnitude. This goes on until the rod begins to radiate violet and ultraviolet. At this point something unexpected begins to happen.

The number of these higher frequency waves is far less than one would expect. The phenomenon is called the ultraviolet catastrophe. It was Max Planck, the same German physicist who told us the higher the frequency of a wave, the greater its energy, who in 1899 came up with an explanation. How he arrived at this explanation is somewhat involved mathematically. He was able to show that the ultraviolet catastrophe could be explained if the energy in the photon is quantized.

When something is quantized, it means that it comes in little individual pieces. Our money is quantized. We can divide a dollar bill into two half-dollars. Each half-dollar can be divided into five dimes and each dime can be divided into two nickels. Finally each nickel can be divided into five pennies, but here we must stop. The penny cannot be broken into smaller units. Our money is quantized in units of the penny.

Planck proved that energy behaves in the same manner. It comes in little indivisible pieces that cannot be subdivided. The smallest unit used to measure energy is the number .00000000000000000000000000000000625 joules which can be written as 6.25×10^{-34} joules and is called Planck's constant. The energy of any photon is found by multiplying Planck's constant by the frequency of the wave. Thus, each photon in red light has 2.5×10^{-19} joules of energy ($6.25 \times 10^{-34} \times 400$ trillion). Each photon in green light has 3.4×10^{-19} joules ($6.25 \times 10^{-34} \times 550$ trillion). Thus, we see mathematically that the higher the frequency, the greater the energy.

Planck's discovery that the energy in photons is quantized marked the birth of what physicists call Quantum Mechanics.

MATTER WAVES AND THE HEISENBERG UNCERTAINTY PRINCIPLE

Imagine an experiment in which small particles are shot from two guns aimed at holes in a wall directly in front of the guns. We would expect the bullets to travel through the holes following a straight line. See Diagram 36A.

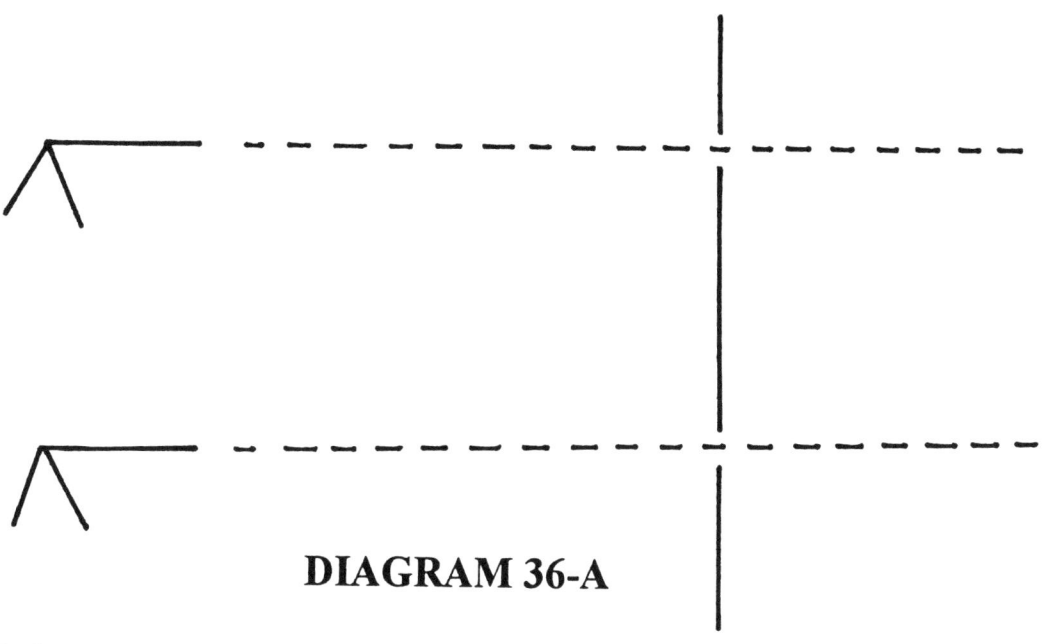

DIAGRAM 36-A

Regular bullets from normal guns would perform just as we would expect; however, when very small bullets are used, we see something quite different. Instead of passing straight through the holes, some of the bullets are deflected upwards and some are deflected downwards. See Diagram 36B.

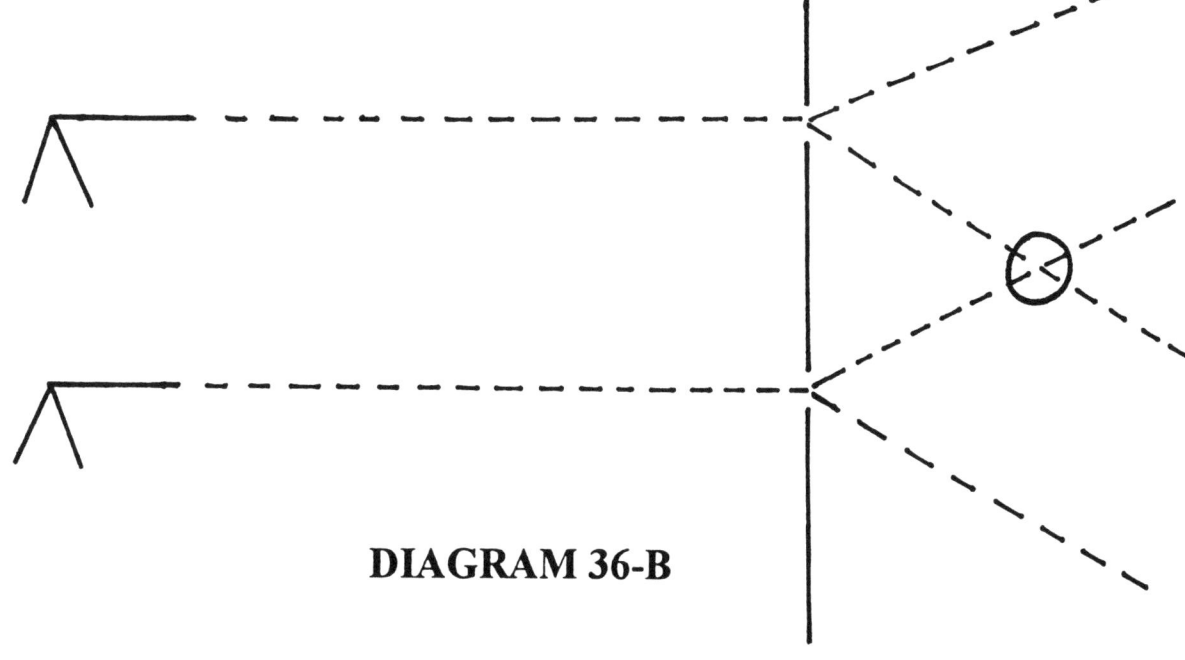

DIAGRAM 36-B

Something else happens which is perhaps even more strange. There is a point where the bullets deflected downward from the top hole and the bullets deflected upward from the bottom hole meet. Here they seem to pass through each other. At this point (circled in Diagram 36B), the bullets can disappear for an instant or they can seem to double up. In other words, the bullets are behaving as though they are waves. It can be shown that all moving objects, including regular bullets, behave like waves; however, the larger the object, the harder it is to detect this property. This is why ordinary bullets from ordinary guns seem to go straight through the holes.

Actually it should not be too surprising to learn that objects behave like waves. If waves can have particles, then why can't particles have waves?

The wave associated with moving objects is named after Louis DeBroglie, the French physicist who discovered them in 1923. DeBroglie waves are sometimes called matter waves. Although they may seem trivial, they have some practical applications. The transistor which is the heart of our computers, TVs, stereos, VCRs etc. operates using the De Broglie wave. There is a barrier inside each transistor that is penetrated by electrons. The only way this penetration can be achieved is by having the electron behave as a wave.

Another practical use of the DeBroglie wave is the electron microscope. The wavelength if the DeBroglie wave is extremely small which makes it very useful for revealing very small objects. Matter waves from moving electrons are used to see extremely small detail in the electron microscope. Perhaps your life, or the life of someone you know, will be saved by a medical discovery made possible by the use of matter waves.

The fact that moving objects can be thought of as waves led Werner Heisenberg (1901-1976) to a famous principle known as the Heisenberg Uncertainty Principle. This principle tells us that it is impossible to measure simultaneously the precise location and the exact speed of an object. We can only give probabilities as to a particle's location. This has been received with great enthusiasm by those concerned with determinism. According to Newtonian physics, once an object's location, mass, and net force applied to it are known, its future positions can be calculated. Applying this to everything in the universe suggests that all events are predetermined. According to the Heisenberg Principle, we cannot pinpoint the exact location of an object; therefore, we cannot predict its exact future. This brings back the concept of a free will.

One very important result of the Heisenberg Principle is that *we can not see anything smaller than the wavelength of the wave used to view the object.* This is way the De Broglie wave of the electron is used in the electron microscope. Its wavelength is much shorter than the wavelength of visible light and thus allows us to see much smaller detail than conventional microscopes that use the longer wavelengths of visible light.

MICHELSON MORLEY EXPERIMENT AND RELATIVITY

The Earth moves in its orbit relative to the sun with a speed of approximately 20 miles per second. This means that if the Earth made a head-on collision with a beam of light, the beam should strike the Earth with a speed twenty miles per second faster than if the Earth were not moving. Light moves with the speed of 186,282 miles per second; thus, common sense suggests that an incoming beam hitting the Earth head-on should approach us at 186,302 miles per second (186,282 + 20).

In 1887 two Cleveland, Ohio physicists, <u>Albert Michelson and Edward Morley</u>, performed an experiment to measure this extra speed of an incoming light wave due to the Earth's movement; however, their instruments failed to show it. Instead, they indicated the speed of the incoming light to be 186,282 miles per second, the same speed they would measure if the Earth were not moving. We find that it does not matter under what conditions we measure the speed of light, it always remains the same. It does not matter whether you are rushing toward the light, away from the light, or standing still; the light will always pass you at 186,282 miles per second. This defies common sense.

It was Albert Einstein who along with the help of others offered an explanation. If the instruments used to measure the speed of a passing light wave could somehow change in size depending on how fast they were traveling, then the speed of light could always be measured the same; in other words, a yardstick appears smaller when it is observed to be moving. When this smaller yardstick is used to measure the speed of a passing light beam, it always indicates a speed of 186,282 miles per second.

The changes in the size of an object are very small for everyday velocities; however, if an object is seen traveling at a speed near the speed of light, the changes become much more significant. In fact, an object seen moving at the speed of light would decrease so that its length would measure zero.

It is important to note that you would not observe these changes in size if you were moving along with the object. This is due to the fact that your measuring devices including your body would also become smaller. You only notice changes when the object is moving <u>relative</u> to you, giving rise to the name <u>Theory of Special Relativity</u>.

Not only does the size of an object appear to decrease as it moves, but the amount of mass in the object increases. This is a paradox. The object becomes smaller, but its mass becomes larger. If the object were to obtain the speed of light, its mass would increase to infinity. It is for this reason that most physicists believe it is impossible to travel faster than the speed of light. To push the infinite mass in an attempt to move it faster than the speed of light would require an infinite force, and there is no such thing as an infinite force.

Perhaps stranger than the change in an object's size and mass is the change in the rate at which time passes. The faster an object moves, the slower time passes. If you were to travel close to the speed of light, time would pass at a very slow rate for you, but again, you would not notice this yourself because clocks moving along with you would tick slower. You might think ten minutes has gone by, while a thousand years has passed for everyone else on Earth. This would be like going ahead into the future, for you could return to the Earth and see what our planet would be like a thousand years from now. Before you attempt this, however, make certain it is what you would want, for there is no way to go back in time.

One might wonder as to why a person in a rocket ship experiences the slowing of time instead of the people back on Earth. Couldn't the person in the ship claim that he or she is standing still and it is the Earth that is moving away with a speed close to the speed of light? This is called the <u>Twin Paradox</u> and can be explained by stating that whoever is accelerated the most will be the younger. If the person in the rocket ship felt a force causing him or her to accelerate to a very high speed, then that person would be the younger. But if we put a huge force on the planet Earth and leave the rocket ship behind, then the people on Earth would all be younger than the person in the rocket.

We have seen that the three fundamental entities of the physical universe are not absolute, but relative. They are dependent on how fast you observe an object to be moving. At ordinary daily speeds, the space occupied by an object, its mass, and the rate of passing time change very, very little. As you approach the speed of light which is always measured to be 186,282 miles per second, then the three entities of space, time, and matter are drastically changed. And if someone saw you moving at the speed of light, that person would observe that time stopped for you, your size reduced to zero, and your mass increased to infinity.

NUCLEAR PHYSICS AND CONSERVATION LAWS

The nucleus of an atom has potential energy because it can do work for us. This energy is nuclear energy. There are two nuclear processes that can be used to harness nuclear energy.

One is <u>nuclear fission</u> which involves breaking apart complex atoms. Uranium is the most complex natural element, having 92 protons and 146 neutrons in its nucleus. Ordinary uranium is called U238, the 238 being the sum of the 92 protons and 146 neutrons. Because all 92 protons are positively charged and since like charges repel each other, the 92 protons are all trying to move away from one another. However, they are held together by the strong interacting force. Occasionally the strong interacting force "lets go" and the atom literally explodes, sending out an assortment of particles. These particles can be detected by using a Geiger Counter. This is why a Geiger Counter is used when prospecting for uranium.

Although U238 does break apart (physicists uses the word decay when talking about an atom that spontaneously explodes), it will break apart much faster if three neutrons are removed from each atom, making it U235. If a large number of U235 atoms are placed together, one atom will spontaneously decay producing fragments that will act as bullets that can jar other U235 atoms causing them to decay. These atoms in turn will produce still more fragments that will cause still more atoms to break apart. This is a chain reaction and in a very short time, trillions of particles are thrown out of the decaying nuclei causing the air to heat to a million degrees. A great fireball is created which produces the familiar mushroom cloud of an atom bomb.

Our nuclear power plants use nuclear fission on a small controlled scale to produce heat which turns water into steam. The steam drives a generator producing electricity. Although nuclear fission is a source of great energy, <u>nuclear fusion</u> is even more potent. Nuclear fusion is almost the opposite of nuclear fission. While nuclear fission begins with complex atoms like uranium and plutonium and involves their breaking apart into smaller fragments, nuclear fusion begins with simple atoms and involves forcing them together to make more sophisticated atoms.

Hydrogen is the simplest atom for it has only one proton in its nucleus. If four hydrogen atoms are squeezed together under tremendous temperature and pressure, they will fuse into one larger atom which is helium. One would expect that a helium atom should have four times the mass of a hydrogen atom. After all, a whole is the sum of its parts and a helium atom is the sum of four hydrogen atoms. And yet, when the mass of the helium atom is measured, its mass is found to be less than the combined mass of four hydrogen atoms. Where is the missing mass?

Nuclear fusion converted the missing mass into a high frequency x-ray or low frequency gamma ray and its corresponding photon. The energy of the photon is found by multiplying the missing mass by the speed of light squared.

$$E = mc^2$$

E represents the energy of the photon

m represents the missing mass

c represents the speed of light

If a large quantity of hydrogen is fused into helium, so many x-rays or gamma rays pass through the air that a much larger fireball than from the atom bomb is produced. This is the hydrogen bomb. In fact, so much heat is produced that instead of just calling this process nuclear fusion, it is often called <u>thermonuclear fusion</u>, thermo meaning heat.

Scientists are currently studying ways to control thermonuclear fusion and no doubt our power plants of the future will be based on it. It surely is the most efficient manner of producing large quantities of energy and, throughout the entire universe, it is the most commonly used method. Our sun and all the other stars shine by thermonuclear fusion.

The last topic we shall consider pertains to the fundamental <u>conservation laws</u> of physics. As new ideas and theories emerge, future physics textbooks will definitely be different from those of today; however, a few fundamental laws of nature will probably never have to be revised. The conservation laws are such basic truths.

We mentioned that energy can not be created or destroyed. This is one of the conservation laws. Positive or negative charge cannot be created or destroyed. This is another conservation law. Still another is the conservation of momentum. <u>Momentum</u> is found in all moving objects and has a tendency to keep the object moving, or if it hits another object it has a tendency to produce motion in the object that is struck. Mathematically, the amount of momentum in a moving object is calculated by multiplying its mass by its velocity. This quantity can never be destroyed, only transferred from one object to another.